Compressors

Compressors

Editor

Manish Bhardwaj

Compressors

Edited by **Manish Bhardwaj**

Printed in 2017

ISBN: 978-1-68117-500-3

Library of Congress Control Number: 2015939256

© 2016 by
SCITUS Academics LLC,
616, Corporate Way, Suite 2, 4766,
Valley Cottage, NY 10989

www.scitusacademics.com

Contents

Preface

Compressors are essential machines for a large number of modern manufacturing processes. Like the hearts pumping life to the production lines, compressors are vital to the operation of key industrial sectors, such as the petrochemical and the mining industries, which rely on compressors for critical tasks, ranging from temperature control to gas transportation and mixing. As a result, there have been continual efforts by the academic and industrial communities to improve the reliability and performance of such turbo machinery as new technologies become available. Stability is a critical factor that limits the performance of compressors. The maximum mass flow output of a compression system is capped by choke, which is generally not a destabilizing phenomenon, and it is caused by the compressed medium reaching sonic conditions.

Editor

Gas Turbine Cogeneration Groups Flexibility to Classical and Alternative Gaseous Fuels Combustion

Ene Barbu[1], Romulus Petcu[1], Valeriu Vilag[1], Valentin Silivestru[1], Tudor Prisecaru[2], Jeni Popescu[1], Cleopatra Cuciumita[1], and Sorin Tomescu[1]

[1]National Research and Development Institute for Gas Turbines COMOTI, Bucharest, Romania
[2]Politehnica University, Bucharest, Romania

INTRODUCTION

The gas turbine installations represent one of the most dynamic fields related to the applicability area and total installed power. The gas turbines have been developed particularly as aviation engines but they find their applicability in many areas, one of which being simultaneously obtaining electric and thermal energy in gas turbine cogeneration plants. The gas turbine cogeneration plants may be classified based on the constructive technology of the gas turbine in [1]: aeroderivative gas turbines plants (up to 10 MW); industrial gas turbines plants, specifically designed for obtaining energy (from 10 up to hundreds MW). An aviation gas turbine with expired flying resource is still functional due to the fact that the flight time is limited as a consequence of the specific safety normatives requirements. Therefore the aeroderivative gas turbine is defined as a gas turbine, derived from an aviation gas turbine, dedicated to ground applications. According to the initial destination, these gas turbines have been designed for maximum efficiency considering the limited fuel quantity available for an aircraft flying large distances. The basic idea in developing the aeroderivative gas turbine has been to transfer all the scientific and technologic knowledge ensuring a high degree of energy utilization (design concepts, materials, technologies, etc.) from aviation to ground [2]. Therefore the obtained gas turbines are lighter, with smaller size, increased reliability, reduced maintenance costs and high efficiency. The remaining resource for ground applications is proportional with the flight resource, being able to reach up to 30,000 hours considering the lower operating regimes. From the point of view of the actual application, the free power turbine groups are the most recommended [3]. Unlike the aeroderivative turbine power units, the industrial power units are built by the original producer with the necessary changes for actual industrial application. The development of aeroderivative and industrial gas turbines has been affected by the progress of the aviation gas turbines in military and civilian fields. Many aeroderivative gas turbines ensure compression rates of 30:1 [4]. The industrial gas turbines are cumbersome but they are more adaptable for long running and allow longer periods between maintenance controls. The base fuel for gas turbine cogeneration groups is the natural gas (with a possible liquid fuel as alternative) but the diversification of the gas turbines users and the increase in fuels price has pushed the large producers to consider alternative solutions. Nowadays the most utilized fuels in gaseous turbines are the liquid and gas ones (classic and alternative). The high temperature of the exhausted gases, approximately 590 °C on some gas turbines, allows the valorization of the heat resulted in a heat recovery steam generator. Due to the fact that the oxygen concentration in the exhausted gases is 11-16% (volume), a supplementary fuel burning may be applied (afterburning) in order to increase the steam flow rate, compared to the case of the heat recovery steam generator [5]. The afterburning leads to an increase in flexibility and global efficiency of the cogeneration group, allowing the possibility to burn a large variety of fuels, both classic and alternative. Nitrogen oxides usually represent the maine source of emissions from gas turbines. The NO_x emissions produced by the afterburning installation of the cogeneration group are different according to the system, but they are usually small and in some cases the installation even contributes to their reduction [6]. The usual methods for NO_x emissions reduction, water or steam injection for flame temperature decrease, affect the gas turbine performances, particularly to high operating regimes, leading to CO emissions increase. It must be noted that the load of the gas turbine also affects the emissions, the gas turbine being designed to operate at high loads. The general theme of the chapter is given by the technological aspects that must be considered when aiming to design a gas turbine cogeneration plant flexible from the points of view of the utilized fuel and the qualitative and quantitative results concerning some classic and alternative gas fuels. Based on the specific literature in the field and the experience of National Research and Development Institute for Gas Turbines COMOTI Bucharest, there are approached theoretic and experimental researches concerning the utilization of natural gas, as classic fuel, and respectively dimethylether (DME), biogas (landfill gas) and syngas, as alternative fuels, in gas turbine cogeneration groups, the interference between flexibility and emissions. It is particularly analysed the issue of reutilization of aviation gas turbines in industrial purposes by their conversion from liquid fuel to gas fuels operation. There is further presented the actual method of conversion for an aviation gas turbine in order to be used in cogeneration groups.

THE AERODERIVATIVE GAS TURBINE – A SOLUTION FOR GAS TURBINE COGENERATIVE GROUPS FLEXIBILITY ON GAS FUELS

The flexibility of the gas turbine cogeneration plants implies reaching an important number of requirements: operating on classic and alternative fuels; capability of fast start; capability to pass easily from full load to partial loads and back; maintaining the efficiency at full load and partial loads; maintaining the emission to a low level even when operating on partial loads. Internationally, many companies with top performance in aviation gas turbines are involved in aeroderivative programs in response to market demands for energy producing installations. The best known among these are: Rolls-Royce, Pratt & Whitney, General Electric, Motor Sich, Turbomeca, MTU, etc. Rolls-Royce has developed the RB 211-H63 gas turbine starting from the aviation RB 211 which, through novel constructive and technologic transformations has been pushed to efficiency up to 41.5%. A 38 MW version will be available in 2013 with the possibility of upgrade to 50 MW in future years [7]. Many gas turbine producers aim to reach the full load in ten minutes from the start. A Japanese project of Mitsubishi Heavy Industries Ltd. (MHI) aims to manufacture a gas turbine operating at 1700 °C inlet temperature and 62 % efficiency. Pratt & Whitney, starting from the PW 100 turboprop, have developed the ST aeroderivative gas turbine family (ST 18, ST 40). The researches conducted at National Research and Development Institute for Gas Turbines COMOTI Bucharest have allowed obtaining aeroderivative gas turbines in the 20 – 2,000 kW range, through valorisation of the aviation gas turbines with exhausted flight resource, obsolete or damaged. Therefore the AI 20 GM (figure 1, right) aeroderivative turboshaft, operating on natural gas, is based on the AI 20 turboprop (figure 1, left). The AI 20 GM is used in power groups driving the backup compressors in the natural gas pumping stations on the main line at SC TRANSGAZ SA. The aeroderivative GTC 1000 (figure 2, right), based on TURMO IV C (figure 2, left), operating on natural gas, is used in a power group driving two serial centrifugal compressors for the compression of the associated drill gas, in one SC OMV PETROM SA oil exploitation, at Țicleni – Gorj. Researches have also been conducted regarding the valorisation of the landfill gas in a aeroderivative gas turbine applicable to cogeneration groups [2]. A project for a cogeneration application using the GTE 2000 aeroderivative gas turbine has been started in 2000. The result of the project is a cogeneration plant, with two independent lines, producing electric and thermal (hot water) energy, located in the municipality of Botosani, with SC TERMICA SA as beneficiary (figure 3, left). The experience acquired from the GTE 2000 cogeneration plant has been used in a new project for a medium power aeroderivative gas turbine cogeneration plant, the application using the ST 18 A aeroderivative gas turbine, manufactured by Pratt & Whitney. The ST 18 A aeroderivative gas turbine has been derived from the aviation PW 100 through redesigning a series of components of which are distinguished the combustion chamber, the case and the intake. Furthermore, the ST 18 A has been designed and manufactured to operate with water injection in the combustion chamber (duplex burners), method that ensures the reduction of NO_x emissions. The application consists in a cogeneration plant, with two independent cogeneration lines, producing electric and thermal (superheated steam used in the oil extraction technologic process) energy. The beneficiary of the application is SC OMV PETROM SA, Suplacu de Barc u, Bihor County (figure 3, right) [8]. What makes the difference between aviation and aeroderivative gas turbines are operating conditions and reliability. Thus, aviation gas turbines have over the period of their useful life so many ordered starts and stops (associated with aircraft flight), short operation between starting and stopping (of hours), short periods between revisions (after each stop) and overhauls (after more than 1,000 hours of operation), the lifespan of about 12,000 cumulative hours of operation. Aeroderivative gas turbines can operate up to 8,000 hours continuously without ordered stop, overhauls are made at intervals up to 30,000 cumulative operating hours and, for some brands, the cumulative operating ranges may be even higher.

Figure 1: AI 20 turboprop (left) and AI 20 GM gas turbine (right) [2].

Figure 2: TURMO IV C turboshaft (left) and GTC 1000 gas turbine (right) [2].

Figure 3: GTE 2000 – Boto ani (left) and 2xST 18 – Suplacu de Barcau (right) plants.

Classic and Alternative Fuels for Gas Turbine Cogeneration Groups

The performances of the gas turbine cogeneration groups (efficiency and emissions) depend in high degree of the type and physical and chemical properties of the used fuels. Depending on the lower heating value (LHV), in relation to natural gas (LHV=30-45 MJ/Nm³), typical gas fuels can be classified as [9]: high heating value (LHV=45-190 MJ/Nm³; butane, propane, refinery off-gas), medium heating value (LHV=11.2-30 MJ/Nm³; weak natural gas, landfill gas, coke oven gas), low heating value (LHV<11.2 MJ/Nm³; BFG - Blast Furnace Gas, refinery gas, petrochemical gas, fuels resulted through gasification etc).

General Requirements Regarding the Utilization of Fuels in Gas Turbines

Figure 4: Control – measurement station for natural gas at 2xST 18 – Suplacu de Barcau plant (left) with booster (right) 1 – cogeneration power plant; 2 – control – measurement station; 3 – booster.

For the gas turbines used in cogeneration groups, for economic reasons, the most used fuels are heavy oil and waste products from various manufacturing or chemistry processes [3]. Using liquid fuels imposes: ensuring combustion without incandescent particles and deposits on the firing tube and the turbine; decreasing the corrosive action of the burned gases caused by the aggressive compounds (sulphur, lead, sodium, vanadium, etc.); solving the pumping and atomization issues (filtration, heating, etc.). A series of fuels must be well purified or filtrated for eliminating water, solid particles or some remiss substances. Heavy liquid fuels must be heated to a convenient temperature to allow their proper pumping and spraying. Coke number and tar number are of particular interest for burning in gas turbines. Coke number (carbon residue) represents the residue left by an oil product (fuel oil, diesel, etc.) when burned in special conditions (closed space, restricted air access, etc.), expressed in mass percent. Tar number indicates the presence of resins, aromatic hydrocarbons, etc. but it must be considered for information only. In order to define the combustion behaviour of a heavy liquid fuel (like oil) it would be indicated to consider as a criterion the product of the coke number and tar number [10]. In terms of reusing aviation gas turbines in industrial purposes, the possibilities of using liquid fuels are decreasing. For each application, the requirements of the beneficiary must be analysed related to the characteristics of the fuels affecting the combustion (density, molecular weight, evaporation limit, flammability temperature, volatility, viscosity, surface tension, latent heat of vaporization, calorific value, the tendency for soot, etc.). In terms of using gas fuels, the problem is less challenging due to their thermal stability, high heating value, lack of soot and tar. However, in order to ensure the pressure level required by the gas turbine, afterburning, etc., the elimination of water and different impurities, a control – measurement station must be provided for the gas fuels to be used (natural gas at 2xST 18 plant – figure 4). Some alternative gas fuels (resulted through gasification and biomass pyrolysis), biogas, residual gases from industrial processes (rich in hydrogen) can play an important role in the operation of the gas turbine cogeneration groups, but they must reach some requirements regarding the calorific value and the composition [11]. Therefore there is necessary to eliminate the impurities, tar, to limit the sulphur and its compounds to 1 mg/Nm^3, respectively the alkaline metal compounds to 0.1 mg/Nm^3 [12].

Alternative Fuels – Characteristics and Consequences regarding their use in Gas Turbine Cogeneration Groups

The biogas produced through anaerobic fermentation is cheap and constitutes a renewable energy source producing, from burning, neutral carbon dioxide (CO_2) and offering the possibility of treatment and recycling for residues and secondary agricultural products, various biowaste, organic waste water from industry, sewage and sewage sludge. The properties and the composition of biogas are different depending on the raw material used, processing system, temperature, etc. The comparative compositions of natural gas and biogas are given in table 1 [13]. For both fuels the main component (giving the energetic value) is the methane (CH_4), the significant differences being given by the high content of CO_2 and H_2S (hydrogen sulphide) in biogas. Technically, the main difference is given by the Wobbe index for natural gas (see chapter 2.2), two times higher than the index for biogas. This leads to a limited possibility of replacing the natural gas with biogas because only gases with similar Wobbe index can substitute each other. The improvement of the biogas can be achieved by replacing CO_2 with CH_4 so as to approach the characteristics of natural gas. Furthermore, the water and hydrogen sulphide must be eliminated to avoid the harmful action of the resulted sulphuric acid on different components of the cogeneration group (gas turbine, afterburning installation, heat recovery steam generator, etc.). Landfill gas resulted from waste deposits represents a cheap energy source, with a composition similar to the biogas resulted from anaerobic fermentation (45-60 % methane, 40-55 % carbon dioxide) [2]. When it comes to using biogas in gas turbine cogeneration groups or introducing it in the natural gas network, special treatment is required (condensate separation, drying, adsorption of volatile substances, etc.). Dimethylether (DME, CH_3-O-CH_3) is a clean alternative fuel which can be produced from fossil fuels, namely coal or vegetal biomass gasification. It can be transported and stored similar to liquefied petroleum gas (LPG), its physical and chemical characteristics, related to natural gas in Ardeal (99.8 % CH_4 and 0.2 % CO_2), being given in table 2 [14]. The flame produced by burning the dimethylether is very similar to the flame produced by the natural gas (figure 5), which makes it suitable to be used as fuel in transportation, cogeneration groups, etc.

   Through biomass of coal gasification (with oxidant agents such as oxygen, air, steam, etc.) it can obtain synthesis gas (syngas) with main components hydrogen (H_2) and carbon monoxide (CO). The syngas can be used to obtain methanol, hydrogen, methane, etc. or can be used as fuel in gas turbine cogeneration groups. Since leaving the gas-producing installation the gas containes ash particles and various compounds of chlorine, fluorine, alkali metals, etc., which must be removed to protect the cogeneration line. Through gasification of different biomass categories and utilization of different gasification technologies, the composition of the resulted gas and the lower heating value (LHV) can vary according to tables 3 and 4 [12, 15]. Tables 1 and 3 show that the lower heating values for biogas and syngas are lower than for the natural gas, requiring, in their application in cogeneration groups, higher mass flow rates with minimum pressure losses. Therefore, the injection nozzles of the gas turbine and the burners of the afterburning installation must be designed for velocities allowing a homogenous mixture between fuel and oxid, as well as low pressure losses. The syngas contains high quantities of hydrogen which affect the combustion in gas turbine cogeneration groups in terms of flame stability, combustion efficiency, etc. Using hydrogen as fuel and introducing a component with dilution role (steam, nitrogen, etc.) the operation of the gas turbine is affected [16].

Table 1. Composition, physical and chemical proprieties for natural gas and biogas [13]

No.	Name	Natural gas	Biogas
1	CH_4 [vol %]	91.0	55-70
2	C_nH_{2n} [vol %]	8.09	0
3	CO_2 [vol %]	0.61	30-45
4	N_2 [vol %]	0.3	0 - 2
5	Lower heating value [MJ/Nm3]	39.2	23.3
6	Density [kg/Nm3]	0.809	1.16

Table 2. Physical and chemical characteristics for natural gas (Ardeal) and dimethylether [14]

No.	Name	Natural gas (Ardeal)	Dimethylether
1	Theoretical combustion temperature [0C]	1,900	2,000
2	Autoignition temperature [0C]	650-750	350
3	Lower heating value [MJ/Nm3]	35.772	59.230

| 4 | Explosion limit [% gas in air] | 5 - 15.4 | 3 - 18.6 |
| 5 | Density [kg/Nm3] | 0.716 | 2.052 |

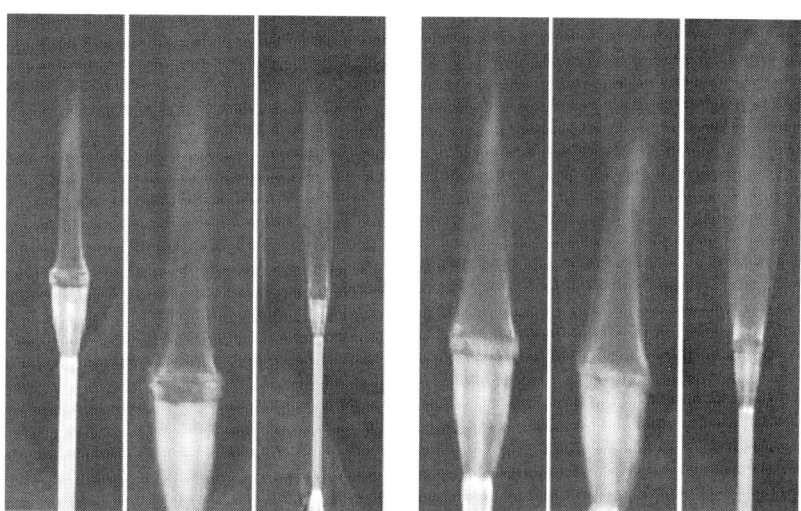

Figure 5: Flame of Bunsen burner, with grid type flame stabilizer, on natural gas (left) and dimethylether (right) [14].

Table 3: Chemical composition of syngas and lower heating values resulted from biomass gasification [15].

Syngas chemical composition [%]							Lower heating value [MJ/Nm3]
Name	CO	H2	CH4	CnH2n	CO2	N2	
Dry oak	18.3	16.9	2.8	0.5	16.0	-	5.422
Dry beech	19.4	17.5	2.6	0.6	15.0	49.3	5.526
Dry fir	15.1	19.1	1.6	0.9	15.8	57.1	4.053
Wood coals	31.2	6.3	2.9	-	2.5	57.1	5.702

Table 4: Chemical composition of syngas and lower heating values resulted from different methods of gasification [12]

CO [%]	H2 [%]	CH4 [%]	N2 [%]	H2O [%]	CO2 [%]	LHV [MJ/Nm3]	Observations
16	6	4	56	18	-	4.1	Air gasification
16	6	4	56	15	3	4.1	Air gasification
40	13	15	3	-	29	11.826	Oxygen gasification

Solving the fuels interchangeability issue for gas turbine cogeneration groups, by developing high level combustion technologies for alternative fuels, particularly hydrogen, will have a major impact on system efficiency and environment.

Fuels Interchangeability and Validation Criteria

Interchangeability in gas turbine cogeneration groups represents the capability to replace a gas fuel with another without affecting the application or the installation burning the gas fuel. The used gas fuels consist in mixtures of combustible gases (methane and other light hydrocarbons, hydrogen, carbon monoxide) and inert gas (mostly nitrogen, carbon dioxide, water vapor). Depending on the combustible gases ratio (usually methane), the gas fuels can have high or low heating value. Density and temperature of the used fuel, as well as the environmental temperature, can affect the performances and lifespan of the equipments in the cogeneration group. According to these influence factors, the most important parameter for characterizing the interchangeability is the Wobbe index (named after engineer and mathematician John Wobbe), defined as ratio between the lower heating value (LHV) and the sqare root of density of the fuel, relative to air density (d_{rel}):

$$Wo = LHV/(d_{rel})^{0.5} \tag{1}$$

$$d_{rel} = \rho_{comb}/\rho_{air} \tag{2}$$

Therefore, two gas fuels, with different chemical compositions but the same Wobbe index, are interchangeable and the heat delivered to the equipment is equivalent for the same fuel pressure. Table 5 gives the values of Wobbe index for several gas fuels. In order to consider the temperature of the fuel, the Wobbe index can be corrected with the temperature. According to [17], two fuels are interchageable if they respect:

$$\frac{\Delta p_2}{\Delta p_1} = \left(\frac{Wo_1}{Wo_2}\right)^2 \left(\frac{A_1}{A_2}\right)^2 \tag{3}$$

where Δp_1 and Δp_2 represent the overpressure of fuel 1, respectivelly 2, Wo_1 and Wo_2 – Wobbe indexes of fuel 1, respectively 2, A_1 and A_2 – injection nozzle area for the two fuels.

Table 5. Wobbe index for various gases [2, 13, 14]

No.	Gas name	Wobbe index [(MJ/Nm3]
1	Natural gas	48.554
2	Liquefied petroleum gas	79.993
3	Methane	47.947
4	Ethane	62.513
5	Propane	74.584
6	Carbon monoxide	12.812
7	Biogas	27.3
8	Dimethylether	47.422
9	Hydrogen	38.3

Therefore, the validation criteria for replacing a fuel with an equivalent one are given by: autoignition temperature, flame temperature (with higher influence on NO_x formation), flame velocity, flashback, efficiency, NO_x and CO emissions, flue gas dew point, etc. Autoignition temperature of gas fuel in mixture with air is the temperature on which the instantaneous and explosive autoignition occurs, without the existence of an incandescent source of ignition. The turbulent flame is generally less stable than the laminar flame, the instability in flame front break-up field being emphasized by the increase in tube diameter. Free swirl turbulent flames are more prone to flame front break-up than the laminar ones due to the higer periferal jet velocity. For turbulence angles greater than 30°, the stability area is achieved on the contour of the burner only for rich mixtures [18]. In areas with poor mixture, due to the decrease in velocity, the backflow can occur without flame attachment on the burner edge. The velocity distribution in the swirl flow determines the stabilisation of the flame as a central suspended one. Components with rapid burning, such as hydrogen, accelerate the flame velocity with a tendency to backflow or extinguishment. The backflow tendency of the flames is proportional with the ignition velocity of the fuel gas, a high velocity leading to a high effect. It is also dependent of the primary air proportion and the components with reduced burning velocity can lead to flame front break-up. In order to consider these factors, an empiric relation has been established for the flame front break-up index at interchangeability I_{ret} [19]:

$$I_{ret} = \frac{k_i f_i}{k_b f_b} \left(\frac{LHV_i}{LHV_b} \right)^{0.5} \tag{4}$$

where: k – constant concerning the flame front break-up limit; f – factor concerning primary air; LHV – lower heating value; b and i – indexes regarding the control fuel, respectively the replacement fuel. A particular issue is raised by the fuels with reduced heating value. Therefore, the landfill gas contains over 40 % CO_2, requiring a suitable fuel feeding in order to achieve combustion. The fuels with reduced heating value have a small range of flammability requiring, at partial loads or transient operating regimes, the utilization of a supplementary fuel (such as propane). The mass flow rates necessary for gas turbine operation on reduced heating value gas fuels are high (neglecting the water or steam injection in the gas turbine) compared with the operation on natural gas, fact that modifies the compressor's operating characteristic [20]. From biomass gasification with air, it is obtained syngas with LHV of 4-6 MJ/Nm³, and from the gasification with steam or oxygen (see table 4) LHV of 9-13MJ/Nm³. An alternative for increasing lower heating value is the mixing with natural gas. Therefore, if the landfill gas has a LHV of 17-20 MJ/Nm³, an equivalent lower heating value can be obtained by mixing 60 % gas with reduced heating value with 40 % CH_4, with respect to the composition described in [21].

Converting the Aviation Gas Turbines from Liquid to Gas Fuels Operation

The complexity of thermo-gas-dynamic processes defining the gas turbine operation in a cogeneration group require theoretical and experimental research activities on gas turbines in order to accomplish the conversion from liquid to gas fuels operation. For the gas turbines on market, in exploitation, the exploitation and maintenance technical specifications are generally known, being provided by the producer. When the object of the research is an existing gas turbine lacking the technical documentation which completely define the contructive solution, the issue must be approached through activities of experimentation, measurements, CAD 3D modelling, numerical simulation in CFD environment, constructive modifications and renewed experimentation in order to validate the constructive solutions, permanently aiming the performances correlated with the maximum effectiveness (thrust, power), minimum specific fuel consumption, maximum efficiency, versatility on fuel conversion, maximum availability, minimum operation and maintenance costs.

General Criteria – Researches Concerning the Modifications on a Gas Turbine for Gas Fuel Operation

The basic procedure for an aeroderivative gas turbine is to keep the rotor assembly, compressor – turbine, which is the „heart" of the gas turbine, form the aviation gas turbine and to redesign the combustion chamber in order to operate on a different fuel than the kerosene. Therefore, for the basic gas turbine in the turboshaft category, at least the combustion chamber must be designed for gas fuels operation. The shaft of the power turbine is mechanicaly connected to a driven load, mechanical work consumer, depending on the application involving the aero-derivative gas turbine (electric generator, compressor, pump, etc.). The command and automatic control system of the aero-derivative gas turbine are designed depending on the application. The bearings can be redesigned, achieving a conversion from rolling bearings to slide bearings. For the basic turboprop (destined for propeller aircrafts), at least the combustion chamber and the reducing gear box and/or the gas generator's turbine must be redesigned, depending if the turboprop does or does not include free turbine. Usually, only the gas generator is used, eliminating the gear box. The issues concerning the automatic control system and the bearings are identical to those of the turboshaft. For the basic turbojet (simple flow jet predominantely for military aircrafts) the redesigning of the combustion chamber and the designing of a power turbine gas-dynamicaly connected to the gas generator are necessary [2]. The issues concerning the automatic control system and the bearings are also identical to those of the turboshaft. Regarding the combustion chamber, is desired to constructively alterate it as little as possible, maybe only in terms of injection system. Due to the fact that the rest of the parameters characterizing the operating process remain unchanged, those regarding zero velocity and ground conditions of the basic gas turbine, the operation of the combustion chamber can be considered as in terms of gas-dynamic similarity. A first problem that must be studied when replacing the fuel is maintaining the combustion efficiency. A second one concerns the maintaining of constructive-functional temperature distribution (on the walls of the firing tube, in the outlet area of the combustion chamber and inlet area of the turbine). On the background of the assembly gas-dyanmic characteristics, the unevenness of the temperatures field on the outlet of the combustion chamber (temperature map) is determined by the geometric characteristics of the dilution area (diameter, length, number and area of holes, etc.) and the characteristics of fuel feeding in the primary area (atomization, jet angles, fuel specifications, etc.). The global temperature map is defined by equation (5) and the radial unevenness for the rotor bladed area is given by equation (6) [3]:

$$\theta_m = \left(T_{max}^* - T_3^*\right) / \left(T_3^* - T_2^*\right) \tag{5}$$

Radial unevenness for the rotor blade area express the manner of operation on the turbine blades:

$$\theta_r = \left(T_{maxr}^* - T_3^*\right) / \left(T_3^* - T_2^*\right) \tag{6}$$

In equations (5) and (6) the significance of symbols is: T*max- maximum temperature peak; T*3- average temperature in the outlet section of the combustion chamber; T*2- average temperature in the outlet section of the compressor; T*maxr- maximum average radial temperature, circumferential arithmetic mean on the entire section. Normal values for θ_m, depending on the gas turbine, are in the 20-25 % range, with reported values of 35 %. In direct connection with the temperature map on the walls of firing tube, the equivalent stress of the material must be considered when replacing a fuel with another. In the case of the AI 24 gas turbine modification for operation on gas fuels in the cogeneration group, a difference of 15 % has been reported in the temperature map, considering the flattening of the temperature peaks when passing through the turbine [22]. The adopted solution has been the generalization of the results obtained by National Research and Development Institute for Gas Turbines COMOTI Bucharest for the AI 20 GM (figure 1) and MK 701 gas generators. In order to achieve the AI 20 GM gas turbine on natural gas (derived from AI 20 on liquid fuel) the adopted constructive solution has been the modification of the injection system, without altering the firing tube (figure 6). The researches for this transformation have been based on test bench experiments with liquid fuel (in low pressure similitude conditions). In order to reach the functional optimum on natural gas, several injection nozzles have been designed and experimented, according to table 6 [3].

Table 6. Configuration of the experimental injection nozzles (see figure 6), for AI 20 GM on natural gas [3]

Nozzle no.	10 Ø3 holes at a 2α angle	Diameter of central hole Ø [mm]
1	900	3
2	700	without central hole
3	800	without central hole
4	700	3
5	800	3
6	1000	without central hole

Only nozzles with 10 holes of the same diameter have been experimented in order to ensure velocity, penetration and safety in operation. The central hole afects the stability of the combustion process, increases the flame radiation and the temperature on the walls of the firing tube. The tie criterion for various injection nozzles for natural gas has been the temperature of the blade on hub. It has been noted that nozzle no. 3 leads to low frequency vibrations in a large range of operating regimes, functionally inadmissible. When operating on natural gas, the combustion efficiency increases with the operating regime, the process being unaffected by the vaporization, but only by the mixing. Following the experimentation, nozzle no. 2 has been selected (with 10 Ø3 holes at 2 =70°, without central hole). For all experimentation regimes, the circumferential temperature map values did not pass 18 %. The same manner of minimum configuration modifications has been applied for the rest of the gas turbines transformed for operating on natural gas (TURMO, MK 701, etc.). Therefore, the firing tube and the combustion chamber case have been kept unmodified for all gas turbines, only redesigning the injection system. Satisfying results have been obtained for the experimentation of TURMO: good stability, but in a more limited range compared with other gas turbines (due to the dependency on the mixing process);

temperature map values of 22 % (for the aimed 20 %). For MK 701, the values on the temperature map have reached max. 20 %. A particular problem is considered when the aim is the integration of the gas turbine, modified for operating on natural gas, with an existing boiler. The heat recovery steam generator can be derived form an energy steam boiler, a technological steam boiler or a hot (warm) water boiler. The integration analysis for an aeroderivative gas turbine with a hot water boiler shows that the temperature of the burned gases on the stack must be in the usual value range and the pressure loss at the passing through the modified boiler (in the cogeneration group) must be lower than the pressure loss on the initial boiler [23]. The modifications necessary for operating the gas turbine on gas fuels with reduced lower heating value, compared with the operation on natural gas, are slightly more complex. Therefore, Mitsubishi Heavy Industries Ltd., with extensive experience in manufacturing gas turbines on BFG (Blast Furnace Gas), considerd the heating value of the gas fuel as the key factor in the modifications scheduled for the gas turbine [24]. Depending on the actual application, more modifications can be operated on the gas turbine, compared with the ones in table 7.

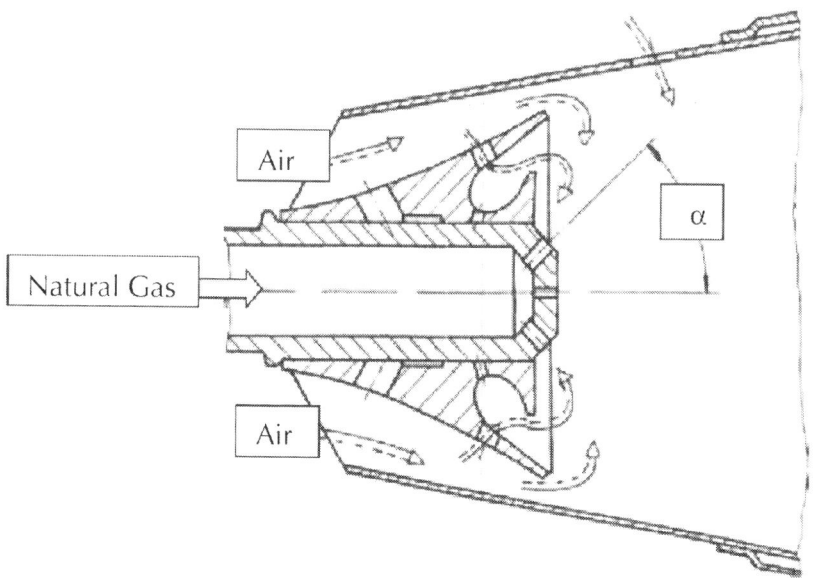

Figure 6: Modification of the injection system for AI 20 GM gas turbine [3].

Table 7. Necessary modifications for a gas turbine, depending on lower heating value of the fuel [24]

Lower heating value [MJ/Nm3]	20.95-41.9 (High)	35.61 (Natural gas)	8.38-29.33 (Medium)	2.51-8.38 (Low)
Air compressor	Standard	Standard	Standard	Modification
Combustor	Standard (Minor mod.)	Standard	Standard (Minor mod.)	Modification
Turbine	Standard	Standard	Standard	Standard
Fuel system	Standard (Minor mod.)	Standard	Standard (Minor mod.)	Modification

Converting a Gas Turbine from Liquid to Gas Fuel Operation for Landfill Gas Valorisation

Converting the gas turbine from liquid fuel to gas fuel operation in order to achieve the valorisation of the landfill gas has known two main steps, respecting the principles in chapter 2.3: converting the TV2-117A gas turbine from operating on liquid fuel (kerosene) to gas fuel (natural gas), resulting the TA2 gas turbine; converting the TA2 from operating on natural gas to operating on landfill gas, resulting TA2 bio. In order to achieve these results, numerous numerical simulations in CFD environment and tests have been used for validating the adopted solutions.

Numerical Simulation, Experimental Activity, Methods and Equipments

Numerical simulation on the TV2-117A gas turbine (figure 7, left) on kerosene has been made in order to obtain a reference model for the gas turbine conversion on gas fuels, particularly landfill gas. An eighth of the geometric model, corresponding to one injection nozzle, has been used in simulations considering the combustion chamber simetry. The boundary conditions have been provided by the producer in the technical specifications for three operating regimes: take-off, nominal and cruise (with the corresponding temperatures of 1123, 1063 and 1023 K). For simulating the combustion process in the TA2 bio gas turbine, the used fuel has been a synthetic landfill gas with equal volume proportions of methane (CH_4) and carbon dioxid (CO_2). The real landfill gas contains other chemical species, in small proportions, which have been considered impurities and have not been taken into account. The numerical simulations have been made on the TA2 with modified injection system, particularly on the injection nozzles level (figure 8). The modelling of the injection nozzles has been achieved starting from the geometry of the natural gas nozzles. Only the injector's outer body have been kept from the liquid operating gas turbine, eliminating all elements related to the atomization system of the liquid fuel. Related to the initial configuration of the injector, only the diameter of the secondary channel and the configuration of the connection with the injection nozzle have been kept unmodified.

Figure 7: TV2-117A gas turbine (left) with detailed combustion chamber area (right).

The numerical simulations for the modified injector (figure 8) have taken into consideration the variation of the injection pressure (7.65 -

8.5 bar), of the injection angle (70 - 85⁰) and the position related to the injector's body L (1 - 5 mm). Following the numerical simulations, the optimum configuration has been selected and the eight injectors have been manufactured along with the injection ramp (figure 10, right), consisting in a circular pipe connected to each injector. The configurations of the injectors for liquid fuel and landfill gas are given in figure 9. The elements eliminated from the initial configuration are the following: the liquid fuel feeding system; the liquid fuel automatic control system; the command system for the actuators controlling the guide vanes and the first three statoric stages of the compressor; the deicing system. The experimentation of TA2 bio has been made in the experimental facility of National Research and Development Institute for Gas Turbines COMOTI Bucharest (figure 10) in the following configuration: TA2 bio gas turbine installed on test bench; test cell lubricating system and fuel feeding system for the gas turbine; exhaust system for the burned gases; monitoring system for acquiring functional parameters. In figure 10 (right) is a pipe ramp ring, yellow color, for gas fuel supply.

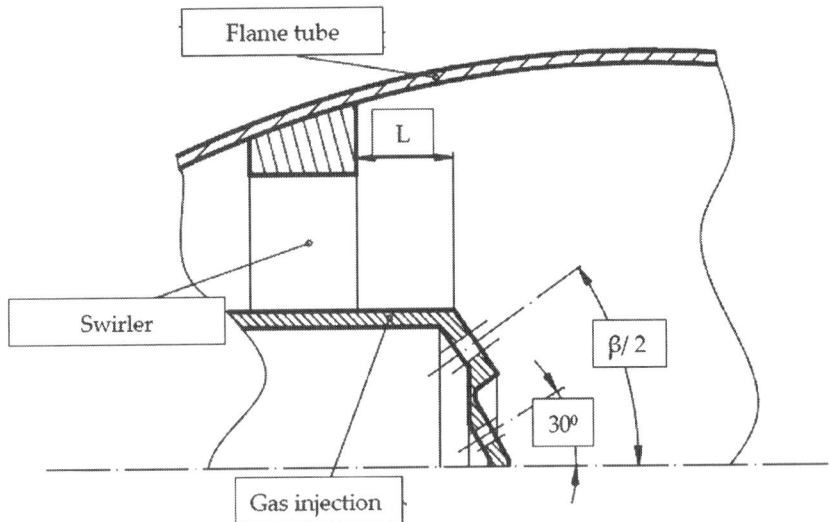

Figure 8: Injection nozzle configuration for landfill gas [2].

Figure 9: Injectors for liquid fuel for TV2-117A (left) and landfill gas for TA2 bio (right) [2].

A series of experimentations have been made, the simulated landfill gas being obtained by mixing natural gas with carbon dioxid (provided from tanks). The measurements have been made with the equipments of the test facilities. A ramp of 17 double thermocouples located at the outlet of the combustion chamber, with measuring points at one third and two thirds of the outer firing tube circumference allow the measurement of the T_{ex} and T_{in} temperatures on two concentric rings (figure 11 right).

Figure 10: TA2 gas turbine (left) and TA2 bio gas turbine in the test cell (centre, right).

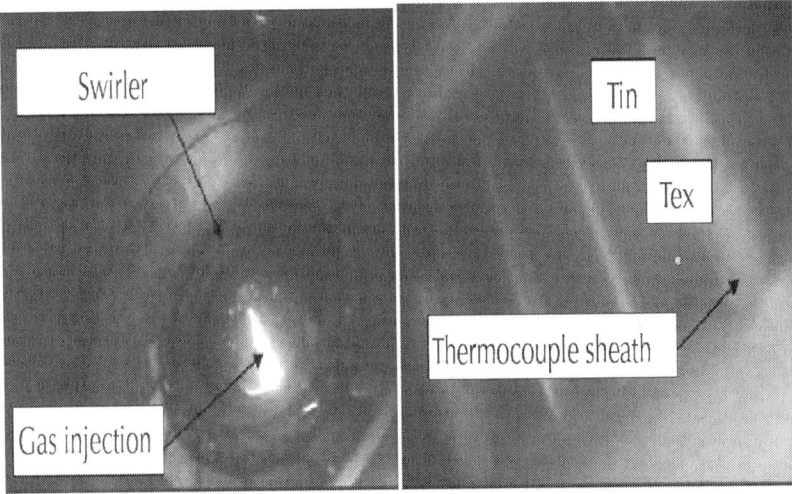

Figure 11: Boroscoping images of the gas injection nozzles – natural gas (left) and the thermocouples (right).

Results and Discussion

The numeric simulations on kerosene [2, 5] have shown that, for the reduced operating regimes, the flame reaches in high degree the area between two adjacent injectors. Table 8 presents the numerical results for landfill gas combustion in terms of methane mass fraction, illustrating the jet shape, and burned gases temperature in the oultlet section of the combustion chamber. Analysis of data in table 8, with respect to temperature maps, aiming to obtain a compact jet in order to protect the walls of the firing tube, have helped selecting the geometric configuration of the injection nozzle: $\beta = 70^0$ and L= 3 mm, used for designing the functional model experimented on TA2 bio, for a mixture of natural gas and carbon dioxide. The experiments have been developed in several series, figure 12 presenting one of the models of variation for the components of the synthetic landfill gas mixture. The experimental results have been synthetized in figures 12 and 13. Figure 13 presents the numerical and experimental results for the outlet section of the combustion chamber.

Table 8. Numerical results for landfill gas combustion simulation [2]

Parameters			Results	
L [mm]	β [0]	p [bar]	Fuel injection jet	Temperature on combustion chamber outlet
1	70	7.65		
3	70	7.65		
5	70	7.65		
3	80	7.65		
3	85	8.50		

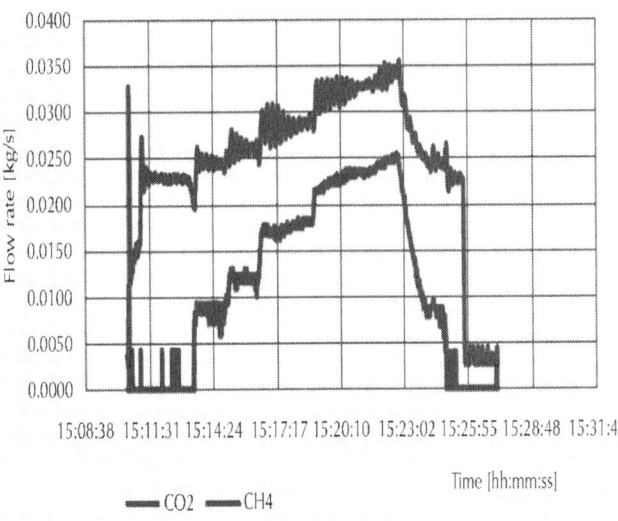

Figure 12: Variation of the mass flow rates of carbon dioxide (CO_2) and natural gas (CH_4) injected in the combustion chamber [2].

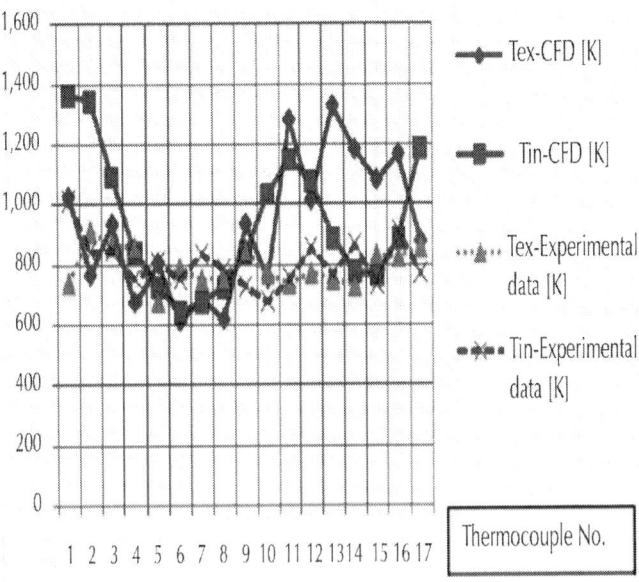

Figure 13: Comparison between the numerical and experimental temperature.

The experimentations have proved a stable operation of the TA2 bio gas turbine on different operating regimes, mainly defined by the mass flow rate and the ratio between the mass flow rate of the natural gas and carbon dioxid. Figure 13, particularly the central area, shows a concordance of the numerical and experimental data, proving that modification of gas turbines operating on alternative gas fuels can be made based on numerical simulations in CFD environment. The model of a cogeneration plant for electric and thermal energy is illustrated in figure 14.

Figure 14: Model of an aeroderivative gas turbine cogeneration plant operating on natural gas and landfill gas.

FLEXIBILITY OF GAS TURBINE COGENERATION GROUPS AND EMISSIONS REDUCTION – FUTURE RESEARCHES

Gas turbine cogeneration groups, alone or in combination with fuel cells, can play an importan role in the general assembly of energy production and emissions reduction. The NO_x reduction must be regarded considering the ensurance of cogeneration group performances in a flexible manner, optimization being possible for a fuel [25]. A higher efficiency implies the optimization of the entire cogeneration plant (gas turbine, afterburning, heat recovery steam generator, etc.). The efficiency must be maintained for partial loads (even below 50 %) or for environmental conditions modification. Starting from 2002, Siemens has taken into consideration the flexibility, eliminating the high pressure barrel of the heat recovery steam generator which requires a long process to reach a certain temperature (in order to avoid the occurence of thermal tensions). Regarding the flexibility, the efficiency and the emissions reduction in gas turbine cogeneration groups, important steps have been made: reduced NO_x burners have been introduced in applications; the lifecycle has been analyzed for efficiency increase; the period between maintenence controls has been extended and the conversion from one fuel to another for multi-fuel engines has improved [7]. The factors determining the formation of pollutant agents exhausted along with the burned gases from the gas turbines are [26]: temperature and air excess coefficient in primary area; homogenization of the process in primary area; residence time of the products; "freezing" characteristic of the reaction near the firing tube, etc. For NO_x reduction, the temperature in the area of the combustion reaction and the areas of maximum temperature and the air jets distribution (stage combustion) need to be reduced. The final configuration of the combustion chamber of a gas turbine is a compromise between the NO_x level, performance and flexibility. Global reduction of the emissions leads to compromises between the emission levels of different components and the assembly characteristics of the combustion chamber (pressure losses, stability and ignition limits, etc.). New concepts must be promoted in order to solve this issue. The usual methods are represented by the water or steam injection in the combustion chamber of the gas turbine, leading to [12]: reduction of NO_x up to 25 ppm (for a 15 % O_2 volume participation in dry burned gases); increase in turbine power due to the increase in fluid mass flow rate (which can compensate the effect of increased temperature during summer); increase of flexibility of the installation in exploitation due to the possibility of load variation through steam flow rate variation. However, the high content of vapours in burned gases can lead to: acid corosion occurence (for fuels containing sulphure); increase in thermal stress on the combustion chamber; reduction of the heat recovery level, etc. Numerical simulations on TV2-117A (figure 15) for water injection in the combustion chamber (through duplex injectors, on natural gas) have shown that the water injection in truncated cone shape, at 45°, characterized by a 12 l/min mass flow rate, leads to minimum NO_x concentration in burned gases of 14 ppm. The analysis of combustion products for TA2 (see chapter 2.4), using NASA CEA program [27], has shown a decrease of the average maximum temperature. The composition of the landfill gas has been considered in equal volume proportions of methane and carbon dioxide, while the composition of the syngas has been considered that given by [19]. The calculation algorythm has started from the stoichiometric reaction of each fuel and imposing the operating regime (in terms of average maximum temperature of 1063 K for nominal regime) in order to determine the minimum quantity of air necessary for the reaction. Obtaining the equilibrium reactions has determined the calculation of the air excess coefficients for each fuel at the given regime, for dry operation. Starting from these initial values, water has been introduced in different proportions, up to 23 %. The supplementary quantity of fuel, necessary to reestablish the operating regime of the gas turbine, in terms of temperature (considering the pressure as unaffected), has been calculated in relation to the quantity of water. The general combustion reactions for each fuel, for the water injection case, for the nominal operating regime, are given by equation (7) for landfill gas andequation (8) for syngas:

$$b\cdot(CH_4 + CO_2) + 2\cdot\lambda\cdot(O_2 + 3.76\,N_2) + a\cdot2\cdot\lambda\cdot H_2O \rightarrow w\,H_2O + x\,CO_2 + y\,N_2 + z\,O_2 \tag{7}$$

$$b\cdot(0.25\cdot CO + 0.09\cdot CO_2 + 0.12\cdot H_2 + 0.52\cdot N_2 + 0.02\cdot CH_4) + 0.225\cdot\lambda\cdot(O_2 + 3.76\,N_2) + a\cdot0.225\cdot\lambda\cdot H_2O \rightarrow w\,H_2O + x\,CO_2 + y\,N_2 + z\,O_2 \tag{8}$$

There have been tracked the thermodynamic of the system and the concentrations of the reaction products, focusing on carbon monoxid (CO) and nitrogen oxides (NO_x). In these conditions, for the two regimes, the calculations have been made up to a injected water coefficient (noted „a") in oxidant of maximum 2, equivalent to 23 % water in oxidant. The maximum proportion of water in oxidant has been limited by the concentration of oxygen resulted from the combustion, minimum 11 %, necessary for the afterburning process. For the nominal operating regime and approximately 15 % water for landfill gas and 12.5 % for syngas, the gas turbine reaches the minimum limit of oxygen.

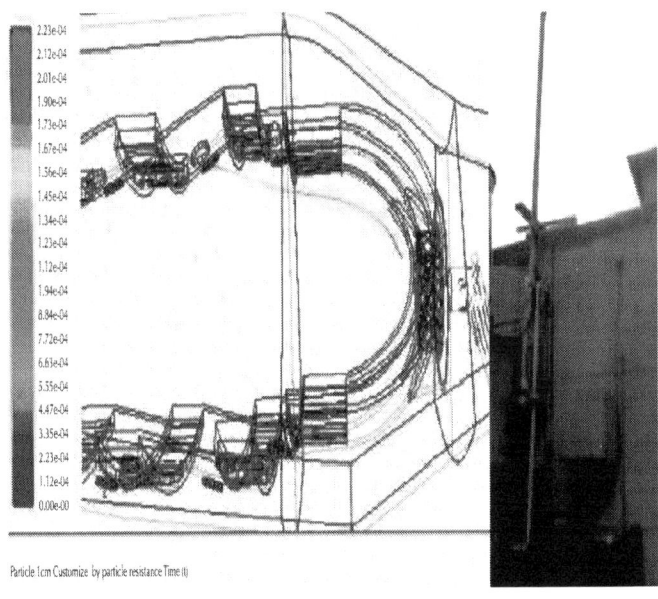

Particle 1cm Customize by particle resistance Time (t)

Figure 15: Numerical simulation of water injection in the combustion chamber of TA2 (left) and atomization tests with the duplex injector (right).

Figure 16 shows the variation of NO_x for the two fuels (landfill gas and syngas) for the nominal regime, depending on the injected water proportion. The results of the calculations illustrate that the use of afterburning along with the operation of the TA2 gas turbine, with water injection, for the good operation of the system, the NO_x produced by the gas turbine at 1063 K can only be reduced to 40 ppm for landfill gas and 38.5 ppm for syngas. The oxygen injected in the air can lead to nitrogen oxides reduction and combustion enhancement resulting [28]: reduction of ignition temperature; increase in flamability limit; increase in adiabatic temperature of the flame; increase in process stability and control; reduction of low heating value fuels consumption, etc. The adiabatic temperature of the flame increases with approximately 50 °C for 1 % increase in oxygen concentration. The volume of burned gases decreases with 12 % for the combustion of natural gas in 3 % oxygen enriched air [29]. Reduction of pollution through combustion in oxygen enriched environment can be used in afterburning installations (for primary or secondary air). Combustion in oxygen enriched environment can increase the efficiency and the flexibility of the cogeneration plant. When adding hydrogen to a gas fuel, there are affected the stability of the flame, the efficiency of the combustion and the emissions. Flame velocity for hydrogen combustion in air, in stoichiometric conditions, reaches 200 cm/s compared to the combustion of methane in air, for which the velocity is approximately 40 cm/s [29]. Adding hydrogen to the gas fuel of the gas turbine or afterburning installation can lead to CO and NO_x emissions reduction.

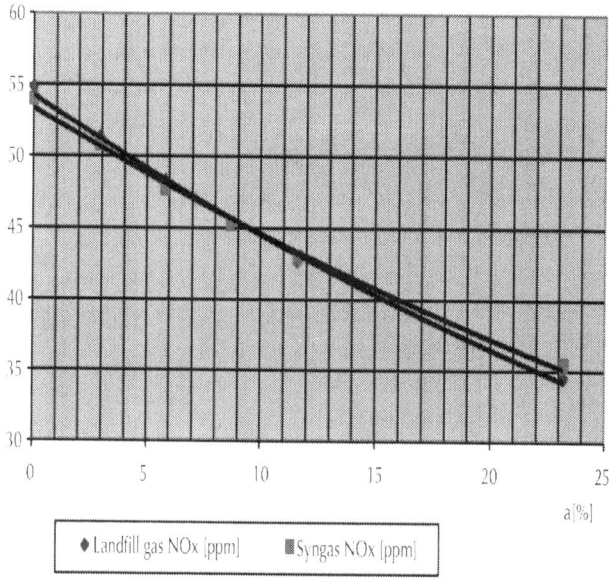

Figure 16: Variation of NO_x concentration for the two fuels, at 1063 K, depending on water proportion in oxidant (a).

Afterburning Installation as Interface between Gas Turbine and Heat Recovery Steam Generator

The burned gases flow when exiting the gas turbine is turbulent and unevenly distributed in transversal section. Therefore, backflow can occur in the transversal section of the recovery boiler. The unevenness of the flow and the variation in burned gases composition affects the operation of the afterburning. Therefore, the afterburning is influenced in terms of efficiency, emissions, flame stability, as well as corrosion of the elements subjected to the action of burned gases. For a good design of the inlet section in the recovery boiler it must be generally considered the following factors [30]: geometry and direction of the gas turbine exhaust; size of heat exchange surfaces; location of the afterburning burner; mass flow rate and average velocity of burned gases exiting the gas turbine; local velocities near the walls and on the first heat exchange surface. The gas turbine exhaust is generally not directly connected with the recovery boiler. After exiting the gas turbine (the case of 2xST 18 Cogeneration Plant at Suplacu de Barcau), the burned gases pass through a silencer, a by-pass assembly, a transom for the connection with the burner and then the afterburning chamber [8]. The gases flow must be parallel with the axis of the burner's connector (perpendicular to the burner plane). A uniform distribution of the flow in the transversal section ensures a good operation of the heat recovery steam generator, particularly regarding the superheater. Therefore, the necessary premises are created for ensuring low emissions on the cogeneration group. If the burned gases or the air are uneven distributed, significant variation of the temperatures downstream the burner can occur. Velocity variation in the transversal section, upstream the burner, must not exceed, on 90 % of the burner's section, ± 15 % of the average velocity measured on the entire transversal section. In reality, the burned gases temperature downstream the burner will never be perfectly uniform. Even for a perfect flow distribution of the turbine gases, upstream the burner, the temperature in the area of each burner module will be higher than the temperature between the modules. Therefore, the infrared analysis of the channel connecting the gas turbine and the afterburning installation (silencer – by-pass assembly – connecting transom), at 2xST 18 Plant, has shown unevenness in temperature distribution (figure 17). Considering these phenomena, the afterburning installation can compensate, in good conditions, the mass flow decrease in burned gases produced by the gas turbine at partial loads, keeping a corresponding load on the heat recovery steam generator. In case of turbine stopping, the heat recovery steam generator with the fresh air afterburning is able to keep the steam production at a certain level.

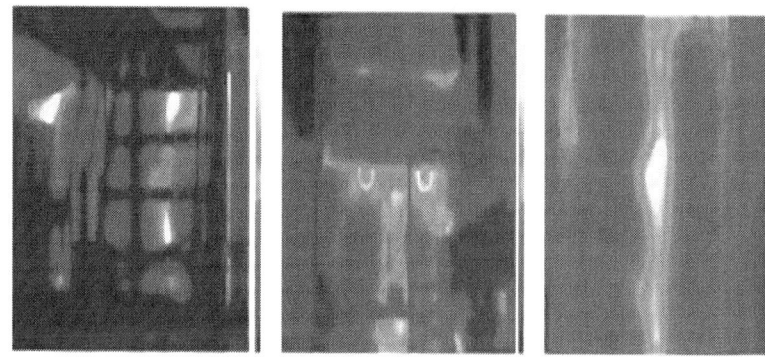

Figure 17: Temperature isotherms, in infrared, in the channel connecting the gas turbine and the afterburning installation (silencer – by-pass assembly – connecting transom).

Future Research

Future research is part of the general context of increasing the flexibility of gas turbine cogeneration groups, the efficiency and reducing the emissions using numerical simulations in CFD environment and experimentations related to: utilization of alternative fuels in gas turbines and afterburning installations, injection of fluids in the cogeneration line in order to reduce the emissions, integrating the gas turbine with fuel cells, etc.

CONCLUSIONS

Along with the flexibility to alternative fuels feeding, the flexibility of a gas turbine cogeneration plant assumes the accomplishment of several requirements: capability of fast start; capability to pass easily from full load to partial loads and back; maintaining the efficiency at full load and partial loads; maintaining the emission to a low level even when operating on partial loads. Using aeroderivative gas turbines in the cogeneration field has allowed the scientific and technologic knowledge transfer utilization (design concepts, materials, technologies, etc.), which ensures a high degree of energy, from aviation to ground applications. The experience of National Research and Development Institute for Gas Turbines COMOTI Bucharest, in the field of aeroderivative gas turbines (AI 20 GM, TURMO, MK 701, etc.) has allowed the conversion of a gas turbine from liquid fuel to landfill gas, for cogeneration, in stable operating conditions.

REFERENCES

1. Cenusa V., Benelmir R., Feidt M., Badea A. On gas turbines and combined cycles. http://www.ati2001.unina.it/newpdf/Sessioni/Macchine/Impianti/03-Cenusa-Benelmir-Feidt-Badea.pdf (accessed June 5, 2012).

2. Petcu R. Contributii teoretice si experimentale la utilizarea gazului de depozit ca sursa de energie. Teza de doctorat - Decizie Senat nr. 100/12.02.2010. Universitatea Politehnica Bucuresti; 2010

3. Carlanescu C. Contributii la problema selectarii si modificarii motoarelor de aviatie pentru utilizarea in scopuri industriale. Teza de doctorat. Universitatea Gheorghe Asachi Iasi; 1994

4. Energy and Environmental Analysis. Technology Characterization: Gas Turbines. http://www.epa.gov/chp/documents/catalog_chptech_gas_turbines.pdf (accessed June 6, 2012).

5. Barbu E., Vilag V., Popescu J., Ionescu S., Ionescu A., Petcu R., Cuciumita C., Cretu M., Vilcu C., Prisecaru T. Afterburning Installation Integration into a Cogeneration Power Plant with Gas Turbine by Numerical and Experimental Analysis. In: Ernesto Benini (ed.), Advances in Gas Turbine Technology. Rijeka: InTech; 2011. p. 139-164. Available from http://www.intechopen.com/articles/show/title/afterburning-installation-integration-into-a-cogeneration-power-plant-with-gas-turbine-by-numerical-(accessed June 6, 2012).

6. Stationary Sources Branch. Stationary Gas Turbines - 40 CFR Part 60. http://www.cdphe.state.co.us/ap/down/statgas.pdf (accessed June 6, 2012).

7. Breeze P. Efficiency versus flexibility: Advances in gas turbine technology. PEI 01/04/2011. http://www.powerengineeringint.com/articles/print/volume-19/issue-3/gas-steam-turbine-directory/efficiency-versus-flexibility-advances-in-gas-turbine-technology.html (accessed May 31, 2012).

8.   Barbu E., Ionescu S., Vilag V., Vilcu C., Popescu J., Ionescu A., Petcu R., Prisecaru T., Pop E., Toma T. Integrated analysis of afterburning in a gas turbine cogenerative power plant on gaseous fuel, WSEAS Transaction on Environment and Development, 2010; 6(6) p. 405-416. http://www.wseas.us/e-library/transactions/environment/2010/89-806.pdf (accessed June 5, 2012).

9.   Jones R., Goldmeer J., Monetti B. Addressing gas turbine fuel flexibility. GE Energy. http://www.ge.com/cn/energy/solutions/s1/GE%20Gas%20Turbine%20Fuel%20Flexibility.pdf (accessed June 6, 2012).

10. Pimsner V., Vasilescu C., Radulescu G. Energetica turbomotoarelor cu ardere interna. Bucuresti, Editura Academiei RSR, 1964

11. Marco Antonio Rosa do Nascimento and Eraldo Cruz dos Santos. Biofuel and Gas Turbine Engines, Advances in Gas Turbine Technology. In: Ernesto Benini (ed.), Advances in Gas Turbine Technology. Rijeka: InTech; 2011. p. 116-138. InTech, Available from: http://www.intechopen.com/books/advances-in-gas-turbine-technology/biofuel-and-gas-turbine-engines (accessed June 6, 2012).

12. Oprea I. Posibilitati de utilizare a gazelor provenite din biomasa in instalatii de turbine cu gaze. ETCN-2005, 30 iunie-1 iulie 2005, Bucuresti, p. 135-139

13. Jensen J., Jensen A. Biogas and natural gas, fuel mixture for the future. 1st World Conference and Exihibition on Biomass and Energy, 2000, Sevilla. Available from http://www.dgc.eu/pdf/Sevilla2000.pdf (accessed June 11, 2012).

14. Panoiu P., Marinescu C., Panoiu N., Oroianu I., Mihaescu L. Posibilitati de utilizare a dimetileterului in scopuri energetice. http://caz.mecen.pub.ro/panoiu.pdf (accessed June 11, 2012).

15. Calin L., Jadaneant M., Romanek A. Gazeificarea biomasei lemnoase. Curierul AGIR, 1-2, ianuarie-iunie 2008, p. 87-90

16. Chiesa P., Lozza G., Mazzocchi L. Using hydrogen as gas turbine fuel, Journal of Gas Turbine and Power, January 2005, vol. 127 73-80 http://www.netl.doe.gov/technologies/coalpower/turbines/refshelf/igcc-h2-sygas/Using%20H2%20as%20a%20GT%20Fuel.pdf (accessed June 12, 2012).

17. Ionel I., Ungureanu C., Popescu F. Analiza nivelului de emisii poluante prin schimbarea combustibilui la cuptoarele de tratament termic. http://www.tehnicainstalatiilor.ro/articole/images/nr12_76-82.pdf (accessed June 14, 2012).

18. Antonescu N., Polizu R., Muntean V., Popescu M. Valorificarea energetica a deseurilor. Bucuresti. Editura Tehnica; 1988

19. Ionel P., Borcea Fl., Barbu E., Marinescu C., Ciobanu C. Mihaescu L. Utilizarea combustibililor gazosi regenerabili pentru producerea de energie.Bucuresti. Editura Perfect; 2008

20. Rainer K. Gas turbine fuel considerations. http://www.scribd.com/doc/76918626/Gas-Turbine-Fuel-Considerations (accessed June 14, 2012).

21. Fossum M., Beyer R. Co-combustion: Biomass fuel gas and natural gas. http://media.godashboard.com/gti/IEA/ieaCofirNOrep.pdf (accessed June 16, 2012).

22. Ene M., Ion C., Salcianu R. Cercetari de transformare a unei camere de ardere pentru functionare cu gaze naturale. In: TURBO '98, 13-15 iulie 1998, Bucuresti, Romania

23. Zubcu V., Zubcu D., Stanciu D., Homulescu V. Instalatie de cogenerare cu componente recuperate, conditii de compatibilitate. in: TURBO '98, 13-15 iulie 1998, Bucuresti, Romania

24. Komori T., Yamagami N., Hara H. Design for blast furnace gas firing gas turbine. http://www.mnes-usa.com/power/news/sec1/pdf/2004_nov_04b.pdf (accessed June 20, 2012).

25. Richards G., McMillian M., Gemmen R., Rogers W., Cully S. Issues for low-emission, fuel-flexible power systems. Progress in Energy and Combustion Science 2001; 27: p. 141–169.

26. Carlanescu C., Manea I., Ion C., Sterie St Turbomotoare – Fenomenologia producerii si controlul noxelor. Bucuresti: Editura Academiei Tehnice Militare; 1998.

27. Zehe, M.J., Gordon, S. & McBride, B.J. (2002), *CAP: A Computer Code for Generating Tabular Thermodynamic Functions from NASA Lewis Coefficients*, NASA Glenn Research Center, NASA TP—2001-210959-REV1, Cleveland, Ohio, U.S.A., http://www.grc.nasa.gov/WWW/CEAWeb/TP-2001-210959-REV1.pdf (accessed June 26, 2012).

28. Corna N., Bertulessi G. The use of oxigen in biomass and waste-to-energy plants: A flexible and effective tool for emission and process control, Third International Symposium on Energy from Biomass and Waste, 8-11 November 2010, Venice, Italy

29. Drnevich R., Meagher J., Papavassiliou V., Raybold T., Stuttaford P., Switzer L., Rosen L. Low NOx emissions in a fuel flexible gas turbine, Issued August 2004, http://www.netl.doe.gov/technologies/coalpower/turbines/refshelf/reports/41892%20 Praxair%20Final%20Report_Low%20NOx%20Fuel%20 Flexible%20Gas%20Turbine.pdf (accessed June 26, 2012).

30. Daiber J., Fluid dynamics of the HRSG gas side, Power, March 2005, p. 58-63 http://www.babcockpower.com/pdf/vpi-45.pdf (accessed June 26, 2012).

Chapter 2

Micro Gas Turbine Engine: A Review

Marco Antônio Rosa do Nascimento[1], Lucilene de Oliveira Rodrigues[1], Eraldo Cruz dos Santos[1], Eli Eber Batista Gomes[1], Fagner Luis Goulart Dias[1], Elkin Iván Gutiérrez Velásques[1], and Rubén Alexis Miranda Carrillo[1]

[1]Federal University of Itajubá – UNIFEI, Brazil

INTRODUCTION

Microturbines are energy generators whose capacity ranges from 15 to 300 kW. Their basic principle comes from open cycle gas turbines, although they present several typical features, such as: variable speed, high speed operation, compact size, simple operability, easy installation, low maintenance, air bearings, low NO_x emissions and usually a recuperator (Hamilton, 2001).

Microturbines came into the automotive market between 1950 and 1970. The first microturbines were based on gas turbine designed to be used in generators of missile launching stations, aircraft and bus engines, among other commercial means of transport. The use of this equipment in the energy market increased between 1980 and 1990, when the demand for distributed generating technologies increased as well (LISS, 1999).

Distributed generation systems may prove more attractive in a competitive market to those seeking to increase reliability and gain independence by self-generating. Manufacturers of gas and liquid-fueled microturbines and advanced turbine systems have bench test results showing that they will either meet or beat current emission goals for nitrogen oxides (NOX) and other pollutants (Hamilton, 2001). Air quality regulation agencies need to account for this technological innovation. Emission control technologies and regulations for distributed generation system are not yet precisely defined. However, control technologies that could reduce emissions from fossil-fueled components of a distributed generation system to levels similar to other traditional fossil-fueled generation equipment are already available.

Combustion processes can result in the formation of significant amounts of nitrogen dioxide (NO_2) and carbon monoxide (CO). Some manufacturers of microturbines have developed advanced combustion technologies to minimize the formation of these pollutants. They have assured low emissions levels from microturbines fueled with gaseous and liquid fuels.

HISTORY

In fact, the technology of microturbines is not new, as researches on this subject can be found since 1970, when the automotive industry viewed the possibility of using microturbines to replace traditional reciprocating piston engines. However, for a variety of reasons, microturbines did not achieve great success in the automotive segment. The first generation of microturbines was based on turbines originally designed for commercial applications in generating electricity for airplanes, buses, and other means of commercial transportation.

The interest in the market for stationary power spread in the mid-1980 and accelerated in the 1990s, with its reuse in the automobile market in hybrid vehicles and when demand for distributed generation increased (Liss, 1999). Currently, the operation of hybrid vehicles through a microturbine connected to an electric motor, have received special attention from some of the major car manufacturers such as Ford, and research centers (Barker, 1997).

In 1978, Allison began a project aimed at the development and construction of generating groups for military applications, driven by small gas turbines. The main results obtained during testing of these generators revealed: reduction in fuel consumption of 180 l/h to 60 l/h, compared with previous models, frequency stability of about 1%, noise levels below 90 dB and the possibility of using different fuels (diesel, gasoline, etc.). In 1981, a batch with 200 generators was delivered to the U.S. Army, and since then, more than 2,000 units have been provided to integrate the system of electricity generation for Patriot missile launchers (Patriot Systems) (Scott, 2000).

The deregulation of the electricity market in the United States began in 1978 when the Power Utility Regulatory Policy Act (PURPA) revolutionized the energy market in the United States, breaking the monopoly of the electricity generation sector, enabling the beginning of the expansion of distributed generation. Since then there has been a significant increase in the proportion of independent generation in the country and, according to a projection made in 1999 by the Gas Research Institute (GRI), this in-house production should reach 35% in 2015 (Gri, 1999).

With a new market structure, i.e., with the possibility of attracting small consumers of energy, microturbines began to be the target of intense research. Already in 1980, under the support of the Gas Research Institute, a program entitled Advanced Energy System (AES) was initiated with a view to develop a small gas turbine, with typical features of aviation turbine, rated at 50 kW and equipped with a heat recovery for a system cogeneration. The program was abandoned around 1990 by the Gas Research Institute, on the grounds of problems with the final cost of the product (Watts, 1999). Since then, the Gas Research Institute began to support new projects in partnership with several companies, such as the Northern Research & Engineering Energy Systems, also supporting the first efforts of Capstone Turbine Corporation (still under the name of its precursor, NoMac Energy Systems) (Gri, 1999).

Some companies in the United States, England and Sweden have recently introduced in the world market commercial units of microturbines. Among these companies are: AlliedSignal, Elliott Energy Systems, Capstone, Ingersoll-Rand Energy Systems & Power Recuperators WorksTM, Turbec, Browman Power and ABB Distributed Generation & Volvo Aero Corporation.

STATE-OF-THE-ART MICROTURBINES

AlliedSignal microturbine has shaft configuration, works with cycle Regenerative open Brayton, its bearings are pneumatic and it has a drive direct current - alternating current (DC/AC) 50/60 Hz (the frequency is reduced from about 1,200 to 50 Hz or 60 Hz) and the compressor and turbine are the radial single stage. The heat transfer efficiency of this stainless steel regenerator is 80-90%. Besides working with diesel oil and natural gas, this microturbine can burn naphtha, methane, propane, gasoline, and synthetic gas. Its noise level is estimated at 65 dB. A commercial prototype of 75 kW was designed for a 30% efficiency and its installed cost is estimated from $ 22,500 to 30,000 (Biasi, 1998).

Elliott Energy Systems (a subsidiary of Elliott Turbomachinery Company) has a manufacturing and assembly unit in Stuart, Florida with a production capacity of 4,000 units per year. According to Richard Sanders, executive vice president of sales and marketing, Elliott has launched two commercial prototypes: a 45 kW microturbine (TA-45model) and another 80 kW (TA-80), and later, a 200 kW microturbine (TA-200). The TA-45 model is rated at 45 kW (Figure 1) at ISO conditions and its main difference from other manufacturers is that it has oil lubricated bearings and a system starting at 24 volts, which, according to Sanders, is unique to microturbines. The TA-80 and TA-200 microturbines models are similar to the TA-45 model. All three can generate electricity in 120/208/240V and can work with different fuels: natural gas, diesel, kerosene, alcohol, gasoline, propane, methanol and ethanol (Biasi, 1998).

The development works of the components has taken the Capstone in the 90's, build and tested a prototype of a 24 kW microturbine in 1994. And in 1996, Capstone made a project consisting of 37 prototypes for field testing. According to Biasi, 1998, Paul Craig, the President of

Capstone Turbine Corporation, expected the 30-kW business model to have a cost of about $ 500/kW (installed microturbine) and a generation cost of $ 45-50/MWh. Figure 2 shows Capstone microturbine, model C65, which is already commercially available.

Figure 1: Elliott Energy Systems Microturbine, TA-45 model.

Figure 2: Capstone microturbine, model C65 (Capstone, 2012).

Four Honeywell Power Systems microturbines of 70 kW each were, until 2001, being tested in the Jamacha Landfill in New Hampshire - United States. The gas produced in the landfills was about 37% methane, carbon dioxide and air. The gas was cooled to about 14 °C to remove moisture and impurities and then compressed to about 550 kPa for the microturbine power. For the first 3 minutes of turbine operation, the fuel feed was carried out with propane. The system operated in parallel and exported electricity to San Diego Gas & Electric. In September 2001, Honeywell decided to stop manufacturing microturbines and uninstalled the four microturbines from the Jamacha Landfill, Figure 3. Until that time, the microturbines operated for 2000 hours, without showing degradation in performance. Then, the microturbines from Honeywell Microturbines were replaced by turbines with the same capacity from Ingersoll-Rand Power Works™, as shown in Figure 4 (Pierce, 2002).

In order to develop a new generation of microturbines, in 1998 ABB Distributed Generation established a 50/50 joint venture with Volvo Aero Corporation. This partnership joined the experience of Volvo gas turbine for hybrid electric vehicles with the experience of ABB in the generation and energy conversion at high frequency. This joint venture resulted in the development of a microturbine for cogeneration. Operating on natural gas, the MT100 microturbine generates 100 kW of electricity and 152 kW of thermal energy (hot water). As other manufacturers of microturbines, the MT100 has a frequency converter that allows the generator to operate at variable speed.

Table 1. brings is a summary of the main features of microturbine leading manufacturers.

Figure 3: Ingersoll-Rand Power WorksTM installed on the Jamacha Landfill - United States.

Figure 4: Prototype Ingersoll-Rand Power Works™ installed on Jamacha Landfill - United States.

Table 1: Technical characteristics of leading microturbine manufacturers

Model	Manufacturers	Power Output	Set	Total Efficiency (LHV)	Pressure Ratio	TET	Nominal Speed
		kW		%		°C	Rpm
-	AlliedSignal	75	A Shaft	30 (HHV)	3.8	871	85,000
TA 45	Elliott Energy System	45	A Shaft	30	-	871	-
TA 80	Elliott Energy System	80	A Shaft	30	-	871	68,000
TA 200	Elliott Energy System	200	A Shaft	30	-	871	43,000
C30	Capstone	30	A Shaft	28		871	96,000
C65	Capstone	65	A Shaft	29		871	85,000
C200 HP	Capstone	200	A Shaft	33		870	45,000
-	Power WorksᵀM	70	Two Shafts	30 (HHV)	3	704	-
MT 100	ABB	100	A Shaft	30	4.5	950	70,000

Microturbines are lower power machines with different applications than larger gas turbines, having typically the following characteristics:

- Variable rotation: the turbine variable speed is between 30,000 and 120,000 rpm depending on the manufacturer;
- High frequency electric alternator: the generator operates with a converter for AC/DC. In addition, the alternator itself is the engine starter;
- Reliability: some microturbines have already reached 25,000 hours of operation (approximately three years) including shutdown and maintenance;
- Simplicity: the generator is placed in the same turbine shaft being relatively easy to be manufactured and maintained. Moreover, it presents a great potential for inexpensive and large scale manufacturing;
- Compact: easy installation and maintenance;
- High noise levels: to reduce noise levels during operation, microturbines require a specific acoustic system;
- Air-cooled bearings: the use of air bearings avoid lubricants contamination by combustion products, prolongs the equipment useful life and reduces maintenance costs;
- Retrieve: microturbine manufacturers generally use heat recovery of exhaust gas to heat the air intake of the combustion chamber, thus achieving a thermal efficiency of 30%.

CONFIGURATION

Microturbines have similar set-up of small, medium and large size gas turbines, as described byNascimento and Santos (2011), i.e., microturbines are formed by an assembly of a compressor, a combustion chamber and a turbine, as shown in the simplified scheme of Figure 5.

State-of-the-art microturbines have markedly improved in the last years. Several microturbines have been developed by manufacturers with different configurations. Their configuration depends on the application, although they usually consist of a single-shaft microturbine, annular combustor, single stage radial flow compressor and expander, and a recuperator or not. The optimum microturbine rotational speeds at typical power ratings are between 60 to 90,000 rpm and pressure ratio of 3 or 4 : 1, in a single stage.

Gas microturbines have the same basic operation principle as open cycle gas turbines (Brayton open cycle). Figure 5 shows the Brayton open cycle. In this cycle the air is compressed by the compressor, going through the combustion chamber where it receives energy from the fuel and thus raising its temperature. Leaving the combustion chamber, the high temperature working fluid is directed to the turbine, where it is expanded by supplying power to the compressor and for the electric generator or other equipment available.

Microturbines are a technology based cycle with or without recuperation. To produce an acceptable efficiency, the heat in the turbine exhaust system must be partially recovered and used to preheat the turbine air supply before it enters the combustor, using an air-to-air heat exchanger called recuperator or regenerator. This allows the net cycle efficiency to be increased to as much as 30% while the average net efficiency of unrecovered microturbines is 17 % (Rodgers et. al., 2001a).

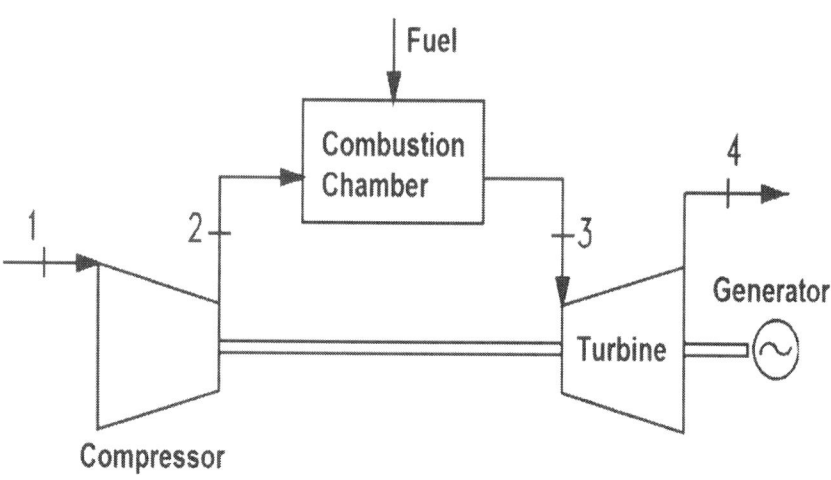

Figure 5: Gas turbine system scheme of a simple open cycle.

As well as in gas turbines, the maximum net power provided by a microturbine is limited by the temperature the material of the turbine can support, associated with the cooling technology and service life required. The two main factors affecting the performance of microturbines are: components efficiency and gases temperature at the turbine inlet.

Furthermore, microturbines usually employ permanent magnet variable-speed alternators generating very high frequency alternating current which must be first rectified and then converted to AC to match the required supply frequency.

Capstone Microturbines, shown in Figure 6, uses a lean premix combustion system to achieve low emissions levels at a full power range. Lean premix operation requires operating at high air-fuel ratio within the primary combustion zone. The large amount of air is thoroughly mixed with fuel before combustion. This premixing of air and fuel enables clean combustion to occur at a relatively low temperature. Injectors control the air-fuel ratio and the air-fuel mixture in the primary zone to ensure that the optimal temperature is achieved for the NO_x minimization. The higher air-fuel ratio results in a lower flame temperature, which leads to lower NO_x levels. In order to achieve low levels of CO and Hydrocarbons simultaneously with low NO_x levels, the air-fuel mixture is retained in the combustion chamber for a relatively long period. This process allows for a more complete combustion of CO and Hydrocarbons (Capstone, 2000).

In addition, the exhaust of microturbines can be used in direct heating or as an air pre-heater for downstream burners, once it has a high concentration of oxygen. Clean burning combustion is the key to both low emissions and highly durable recuperator designs.

The most effective fuel to minimize emissions is clearly natural gas. Natural gas is also the fuel choice for small businesses. Usually the natural gas requires compression to the ambient pressure at the compressor inlet of the microturbine. The compressor outlet pressure is nominally three to four atmospheres.

Capstone microturbine control and power electronic systems allow for different operation modes, such as: grid connect, stand-alone, dual mode and multiple units for potentially enhanced reliability, operating with gas, liquid fuels and biogas. In grid connect, the system follows the voltage and the frequency from the grid. Grid connect applications include base load, peak shaving and load following. One of the key aspects of a grid connect system is that the synchronization and the protective relay functions required to reliably and safely interconnect with the grid can be integrated directly into the microturbine control and power electronic systems. This capability eliminates the need for very expensive and cumbersome external equipment needed in conventional generation technologies (Rodgers et. al., 2001a). In the stand-alone mode, the system behaves as an independent voltage source and supplies the current demanded by the load. Capstone microturbine when equipped with the stand-alone option includes a large battery used for unassisted black start of the turbine engine and for transient electrical load management.

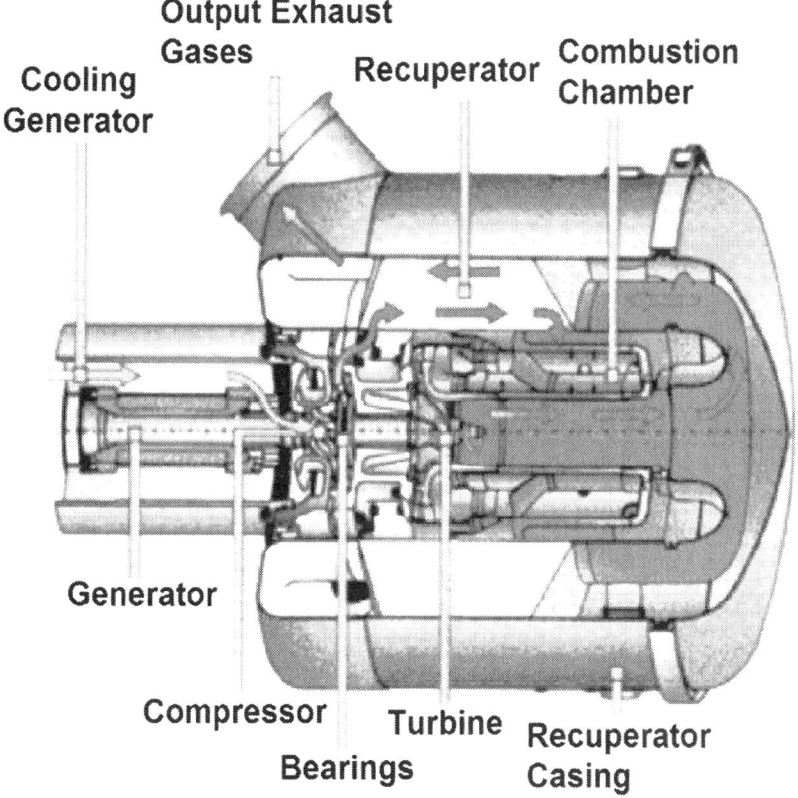

Figure 6: Parts of a Capstone microturbine.

In both operational mode, that is, the grid connect and the stand-alone, the microturbine can also be designed to automatically switch between these two modes. This type of functionality is extremely useful in a wide variety of applications, and is commonly referred to as dual mode operation. Besides, the microturbines can be configured to operate in parallel with other distributed generation systems in order to obtain a larger power generation system. This capability can be built directly into the system and does not require the use of any external synchronizing equipment.

Some microturbines can operate with different fuels. The flexibility and the adaptability enabled by digital control software allow this to happen with no significant changes to the hardware. Power generation systems create large amounts of heat in the process of converting fuel into electricity. For the average utility-size power plant, more than two-thirds of the energy content of the input fuel is converted into heat. Conventional power plants discard this waste heat, however, distributed generation technologies, due to their load-appropriate size and sitting, enable this heat to be recovered. Cogeneration systems can produce heat and electricity at or near the load side. Cogeneration plants usually have up to 85% of efficiency and operation cost lower than other applications. Small cogeneration systems usually use reciprocating engines although microturbines have showed to be a good option for this application. The hot exhaust gas from microturbines is available for cogeneration applications. Recovered heat can be used for hot water heating or low-pressure steam applications.

EXPERIMENTAL SET-UP FOR MICROTURBINE

To perform tests in microturbines, a test bench was built in the Laboratory of Gas Turbines and Gasification of the Institute of Mechanical Engineering, Federal University of Itajubá - IEM/UNIFEI. This bench was composed of a 30 kW regenerative cycle diesel single shaft gas microturbine engine with annular combustion chamber and radial turbomachineries, as shown in Figure 7, and was configured to operate with liquid fuel.

The microturbine engine was tested while in operation with automotive ethanol and pure diesel, respectively. Thermal and electrical parameters, such as mass flows, temperature, composition of exhaust gases and generated power were constantly measured during the tests.

Figure 8 shows the scheme of the microturbine with the measuring points. The microturbine engine was tested during operation with ethanol and diesel at steady state condition and at partial, medium and full loads.

As can be seen in Figure 8, all parameters assessed, during laboratory tests, were acquired and post-processed in a supervisory system developed in the laboratory UNIFEI.

In order to establish whether the fuels were able to feed the engine without presenting any problems regarding the fuel injection system, the kinematic viscosity of each fuel was measured. The composition of the emission gases and the thermal variables were also measured at medium and full loads for each fuel, and their results are presented below. All tests were performed in the grid connection mode.

Figure 7: Capstone microturbine in the laboratory at UNIFEI.

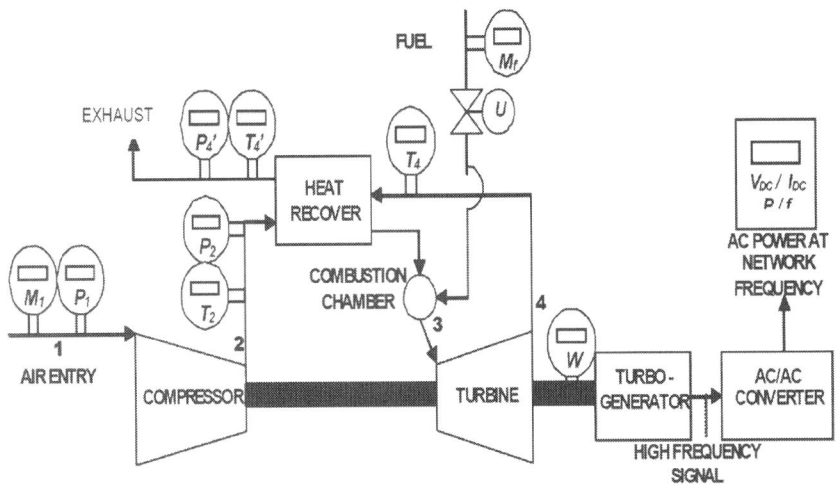

Figure 8: Schematic representation of the test rig and the data acquisition system.

This microturbine is mainly used for primer power generation or emergency and can work with a variety of liquid fuels. This microturbine uses a recovery cycle to improve its efficiency during operation, due to a relatively low pressure, what facilitates the use of a single shaft radial compression and expansion [Cohen, *et. al.*, (1996), Capstone, (2001), Roger, *et. al.*, (2001b), Bolszo (2009)]. Table 2 shows the engine design characteristics at ISO condition.

Table 2: Engine Performance data at ISO Condition

Fuel Pressure	350 kPa
Power Output	29 kW NET (± 1)
Thermal Efficiency	26% (± 2)
Fuel HHV	45,144 kJ/kg
Fuel Flow	12 l/h
Exhaust Temperature	260 °C
Inlet Air Flow	16 Nm³/min
Rotational Speed	96000 rpm
Pressure Ratio	4

For tracking and measuring the tests parameters a type of supervisory software was used in the test bench (given by the turbine manufacturer) along with the data acquisition and the post processing obtained during the tests.

The composition of the exhaust gases was measured in real time using an Ecoline 6000 gas analyzer, reporting the concentration of O_2, CO_2 and hydrocarbons (HC) in volume percentage (%v/v) and NO, CO, NO_2 and SO_2 (ppm) (Sierra, 2008). The fuel high heating value (HHV) was determined by a C-2000 IKA WORKS calorimeter. The accuracy, range and resolution of each instrument used during the tests are shown in Table 3.

Table 3: Accuracy of the measuring instruments

Instrument		Range	Resolution	Accuracy
Fuel Flow		0-100 (l/h)	1.0 (ml)	±1.0 (%) scale
Temperature		0-350 (°C)	0.31 (°C)	±0.8 (%) scale
Pressure		0-10 (bar)	0.01 (bar)	±1.0 (%) scale
Power		0-45 (kW)	0.05 (kW)	±0.5 (%) scale
Calorimeter		--	--	±0.5 (%)
Gas Analyzer	CO (ppm)	0 - 20000	1	± 10 < 300 ± 4 (%) rdg < 2000 ± 10 (%) rdg "/> 2000
	NOx (ppm)	0 - 4000	1	± 5 < 100 ± 4 (%) rdg < 3000

Adjustments to the Microturbine

Due to impurities in the gas or fuel, for instance, in the synthesis or biofuel, a redesign of the gas turbine combustor was necessary. For each type of fuel, a different kind of optimization was needed, in relation to the fuel low heating value (LHV).

To compensate for the lower heating value (LHV) of fuel gases, the fuel injection system must provide a much higher fuel rate than when operating with high heating values. Due to the high rate of mass flow of gas with LHV, the passage of fuel has a much larger cross section than the section corresponding to natural gas. Fuel pipes, control valves and stop valves have larger diameters and shall be designed to include an additional fuel blend, which consists of the final mixture of the recovered gas with natural gas and steam. The pressure drops and the size of the air spiral entering the flame tube must be adjusted to optimize the combustion process. The system must have high safety standards, so the flanges and the gaskets of the combustor and its connections must be safely welded. The system for low LHV must include:

- Fuel line for a LHV;
- Natural gas line;
- Steam line to reduce NO_x;
- Line blending of fuel for LHV;
- Line of nitrogen to purge;
- Lines pilot;
- Compressor;
- Combustion Chamber.

For safety reasons, the loading of the gas turbine to the rated load is accomplished through the use of the fuel reserve. The procedure for replacing the fuel reserve to the main tank is done automatically.

Tests on Gas Turbine Using Liquid Fuel

The performance of a gas turbine is related to the local conditions of the installation and the environment, where pressure and temperature conditions are of great importance.

Due to the diesel low solubility at low temperature, tests with ethanol were performed without premix, and without the use of additives, which increased the cost of fuel.

According to the measuring methodology to be adopted to test gas turbines operating on liquids fuels, the physical-chemical properties of ethanol and diesel are shown in Table 4.

Table 4 also shows the fuel requirements established by the manufacturer of the tested gas turbine along with ASTM D6751 standard specifications for the testing of thermal performance. Regarding emissions a standard ISO 11042-1:1996 was used (NWAFOR, 2004).

Table 4: Ethanol and diesel physical-chemical characteristic

Properties	Ethanol	Diesel	Fuel Limits	ASTM D6751
Sulfur (% mass)	0	0.20	0.05 <	< 0.05
Kinematic Viscosity @40 °C (mm²/s)	1.08	1.54	1.9 – 4.1	1.9 – 6
Density @ 25 °C (g/cm³)	0.786	0.838	0.75 – 0.95	-
Flash Point (°C)	13	60	38 - 66	"/> 130
Water (% Volume)	0.05	0.05	0.05	0.05
LHV (kJ/kg)	23,985.00	42,179.27		

The experimental determination of the ethanol heating value, kinematic viscosity and density were carried out according to ISO 1928-1976 and ASTM D1989-91 standards (ASME, 1997).

The use of different fuels implies the need of mass flow rate adjustments, according to its LHV and density, as without these adjustments, once established a load, the supply system would feed a quantity of fuel depending on the characteristics of the standard fuel (diesel). If the LHV of the new fuel is lower than standard, the gas turbine power could not reach the required demand.

Initially, the engine operated with conventional diesel fuel for a period of 20 minutes to reach a steady state condition for a load of 10 kW. After 20 minutes, the mass flow rates were changed to the fuel corresponding values. At this stage the fuel started to be replaced in order to increase the content of ethanol, by closing the diesel inlet valve and opening the ethanol valve. In order to ensure that all existing diesel power on the engine internal circuitry would be consumed, the engine was left running for 10 minutes with the same load operation, that is, 10 kW.

In order to check if the fuels were able to supply the engine, without causing problems to the fuel injection system, the kinematic viscosity of each fuel was measured. The composition of gas emissions and thermal parameters were also measured in total and average load for each fuel. This whole procedure was performed for the engine operating with loads of 5, 10, 15, 20, 25 and 30 kW in a grid connection mode.

Afterwards the emissions were measured with a gas analyzer, and the load of 5 kW increased. Ten minutes were necessary until it reached steady state again. Exhaust emissions were measured from the exhaust gases and, as mentioned before, the thermal performance data were stored in a personal computer (PC) unit coupled with a PLC (Programmable Logic Controller) data acquisition system, which carried out the data reading at every second.

When tests with ethanol were over, the engine was left running, in order to accomplish the purging of the remaining fuel. After that the engine was once again operated with diesel for ten minutes, and then disconnected and stopped.

PERFORMANCE EVALUATION

The performance showed in this study was obtained from experimental tests at the Gas Turbine Laboratory of the Federal University of Itajubá (GOMES, 2002). Both natural gas and liquid fuel Capstone microturbines and their respective fuel supplying and electrical connection systems were installed and a property measurement was used to obtain the behavior of microturbines operating at partial and full load.

Natural Gas

The microturbine tested on natural gas was a Capstone 330 High Pressure. Table 5 gives the technical information of this machine and the features of the natural gas used in the tests. The natural gas microturbine was tested on the stand-alone mode supplying a resistive load. These microturbines can record operational parameters (temperatures, pressures, fuel usage, turbine speed, internal voltages/currents, status, and many others). Such data can be accessed with a computer or modem connected to an RS-232 port on the microturbine. To supplement these data, additional instrumentation was installed for the tests.

Table 5: General conditions of the analysis

CAPSTONE Microturbine Features		
Model	330 (High Pressure)	
Full-Load Power (ISO Conditions)	30 kW	
Fuel	Natural Gas	
Fuel Pressure	358 – 379 kPa	
Fuel Flow*	12 m^3/h	
Efficiency (LHV)*	27%	
Proprieties of Natural Gas (20 °C and 1 atm)		
Specific Mass	0.6165	
Low Heat Value	36,145	kJ/^{m}3
High Heat Value	40,025	kJ/^{m}3
Ambient Conditions		
Elevation	800	meters
Average Temperature	30	°C

A large battery started the microturbine when disconnected from the grid, preventing any sudden load increase or decrease in the electrical buffer during the stand-alone operation (Capstone, 2001). The start-up took about 2 minutes and the speed was increased from 0 (zero) to 45,000 rpm, occasion when the microturbine started generating electricity. The rotating components of the microturbine were mounted on a single shaft supported by air bearings and a spin at up 96,000 rpm. Figure 9 shows the speed behavior with the microturbine power output.

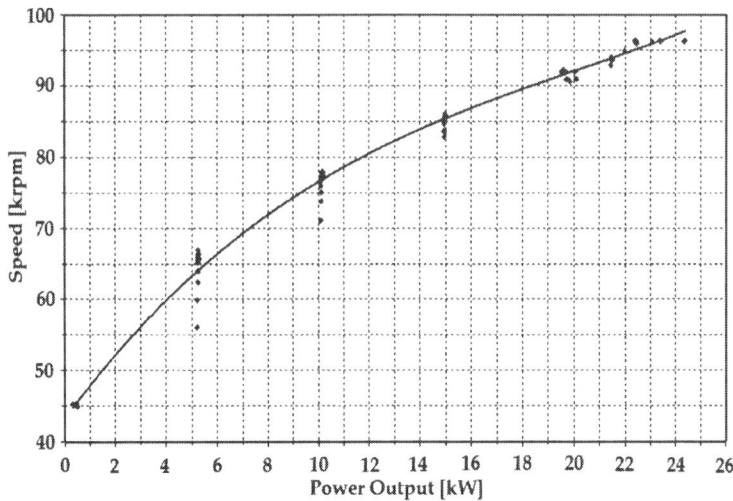

Figure 9: Microturbine speed at partial loads.

Capstone microturbine includes a recuperator which allows the microturbine efficiency to be improved.Figure 10 and 11 show respectively, the exhaust temperature and the efficiency behavior at partial loads. 27 % efficiency is possible at full load.

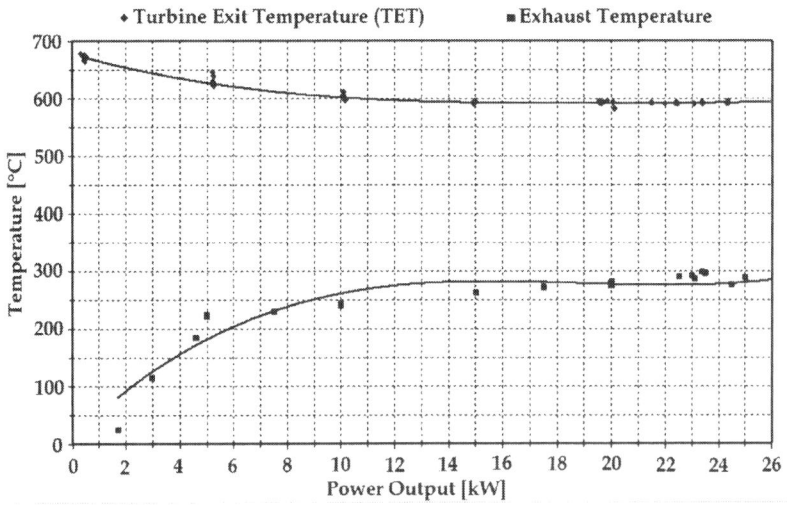

Figure 10: Microturbines exhaust temperature at partial loads.

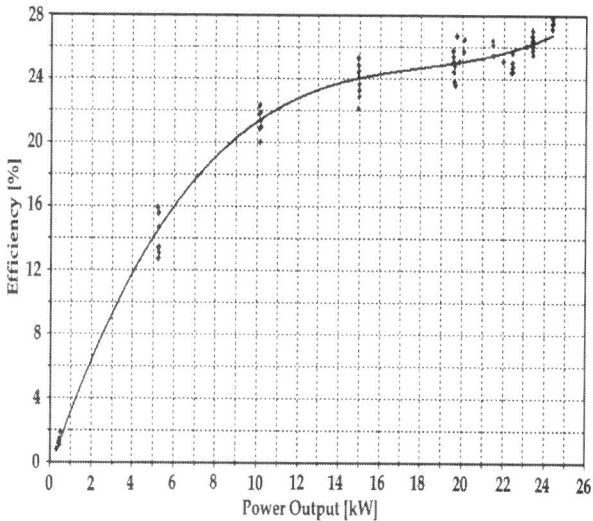

Figure 11: Microturbine efficiency at partial loads.

Figure 12 shows CO and NO$_x$ emissions behavior of a Capstone natural gas microturbine. Combustion occurs in three different steps. The first step is from start-up to about 5 kW. At this step CO formation decreases and emissions of NO$_x$ increase quickly.

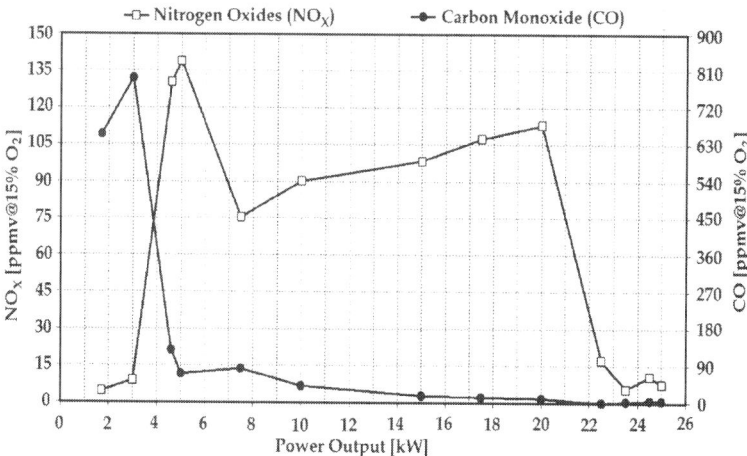

Figure 12: CO and NO$_x$ emissions of a natural gas microturbine at partial loads.

The second step is between 5 and 20 kW, as shown in Figure 12. In the second step the CO formation decreases continuously while emissions of NO_x decrease at first, though increasing but it returns to increase softly slightly up to 113 ppmv. The last step begins at this point. At this step the lean-premix combustion occurs and the NO_x formation diminishes to 5 ppmv.

Emissions of CO_2 depend on the fuel type and the system efficiency. Figure 13 shows CO_2 emissions of a Capstone natural gas microturbine.

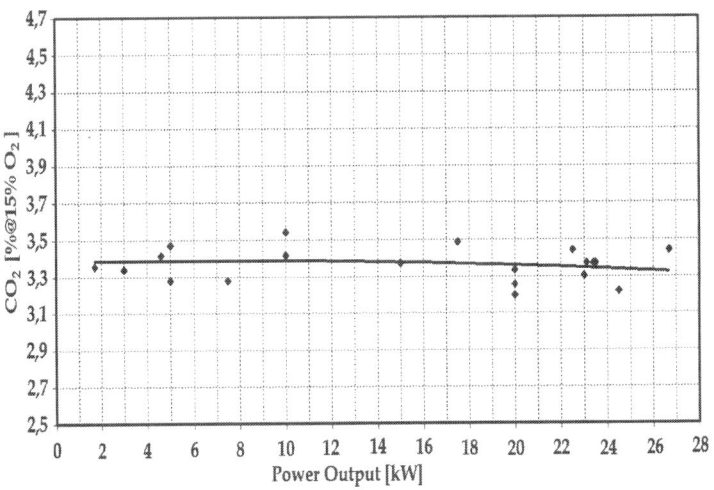

Figure 13: CO_2 emissions of a natural gas microturbine at partial loads.

Liquid Fuel

The microturbine tested on diesel was a Capstone 330 Liquid Fuel. Table 6 gives the technical information of this machine and the features of the diesel used in the tests.

Table 6: General conditions of the analysis

CAPSTONE Microturbine Features	
Model	330 (Liquid Fuel)
Full-Load Power*	29 kW

Fuel	Diesel #2 (ASTM D975)		
Fuel Pressure	35 – 70 kPa		
Fuel Flow*	12.5 l/h		
Efficiency (LHV)*	26%		
Proprieties of Liquid Fuel (20 °C and 1 atm)			
Specific Mass	0.848		
Low Heat Value	42,923	kJ/kg	
High Heat Value	45,810	kJ/kg	
Ambient Conditions			
Elevation	800	meters	
Average Temperature	30	°C	

[i] - * ISO Conditions

The liquid fuel microturbine was tested on the grid connect mode. These data can be accessed with a computer or modem connected to an RS-232 port on the microturbine. To supplement these data, additional instrumentation was installed for the tests. Figure 14 shows the turbine exit temperature and the exhaust temperature at partial loads. These temperatures are before and after the recuperator were used and their difference ranges from 300 to 450 °C.

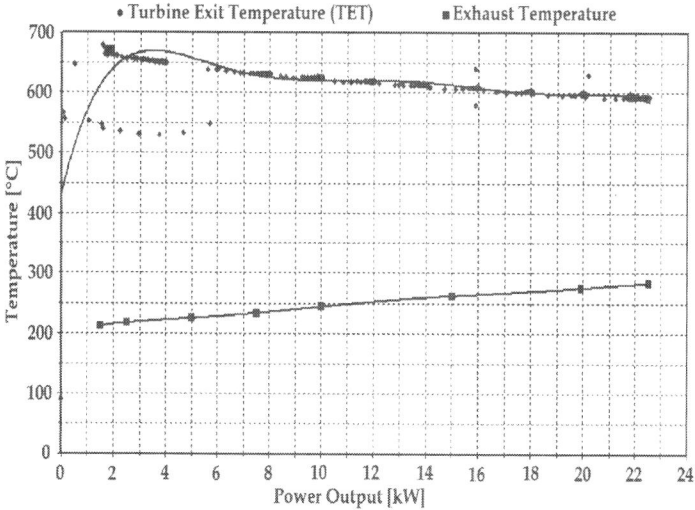

Figure 14: Microturbine exit and exhaust temperature at partial loads.

Figure 15 shows the liquid fuel microturbine efficiency at partial loads. Up to 24.5 % efficiency is possible at full load while the microturbine efficiency is at its highest when Capstone microturbines operate over an output range between 12 kW and full load.

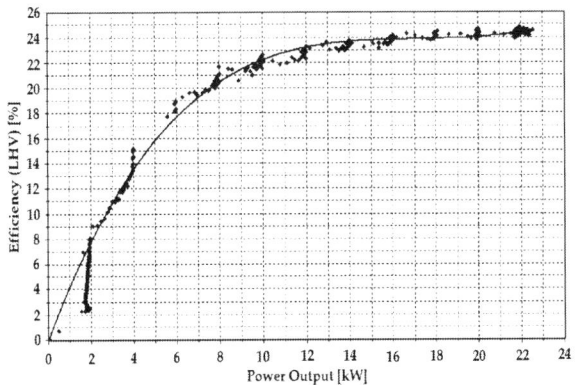

Figure 15: Microturbine efficiency at partial loads.

Figure 16 shows the CO and NO_x emissions behavior of a Capstone liquid fuel microturbine. The CO formation decreases, whereas emissions of NO_x increase as the power output increases due to a rise in the flame temperature.

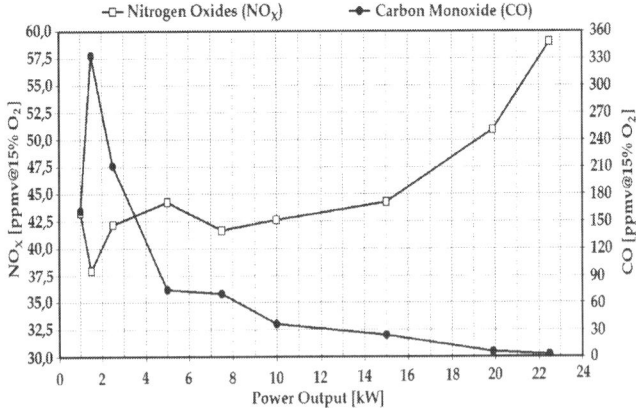

Figure 16: CO and NOX emissions from liquid fuel microturbine at partial loads.

Figure 17 shows the CO_2 and SO_2 emissions of a Capstone liquid fuel microturbine. The emissions depend considerably on the liquid fuel features. While SO_2 emissions are an important emission category for traditional electric utility companies, they are expected to be negligible for distributed generation technologies.

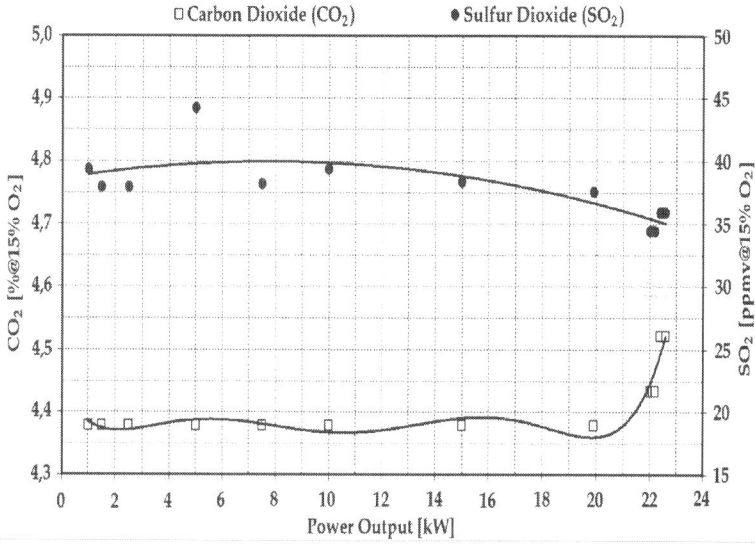

Figure 17: CO_2 and SO_2 emissions from liquid fuel microturbine at partial loads.

MICROTURBINES AND INTERNAL COMBUSTION ENGINE´S EMISSIONS

Table 7 and 8 compare emissions data from internal combustion engines and microturbines. In the absence of a post combustion device, such as a catalytic converter, reciprocating engines can have very high emission levels. Emission levels of microturbines are lower than levels of internal combustion engines as microturbines combustion is a continuous process which allows for a complete burning.

Table 7: NO_x emissions of internal combustion engines (ICE) (Weston, *et. al.*, 2001)

		ICE Natural Gas Without Control	ICE Natural Gas SCR	ICE Diesel Without Control	ICE Diesel SCR
Efficiency	% (HHV)	36%	29%	38%	38%
Nominal Power	kW	1,000	1,000	1,000	1,000
NO_x(@15%O_2)	g/MWh	998	227	9,888	2,132

[i] - SCR: Selective Catalytic Reduction.

Table 8: CO and NO_x emissions of Capstone microturbines

FUEL		**Natural Gas**	**Diesel**
Efficiency*	% (LHV)	27	26
Nominal Power*	kW	30	29
CO (@15%O_2)**	g/MWh	210	80
NO_x(@15%O_2)**	g/MWh	520	280

[i] - * ISO Conditions; ** On Site Conditions (See Table 1)

CASE STUDIES UNDER BRAZILIAN CONDITIONS

Due to the Brazilian governmental incentive to develop the gas industry, the feasibility of many natural gas applications has been doubted. Consequently, the demand for efficiently and environmentally friendly power generation technologies has increased. Many electricity consumers are considering producing their own electricity (Gomes, 2002).

This study analyses the possibility of natural gas application with Capstone microturbines in three cases of power generation: peak shaving in a small industry, base load in a gas station and a cogeneration system supplying buildings in a residential segment

Nowadays it is a trend on microturbines market to reduce investments. This paper analyses the influence of the investment cost of microturbines on the feasibility and cost of the generated electricity, being the cost of fuel a significant part of the electricity final price. The feasibility and the cost of the electricity generated with fuel were also assessed. This study used electric energy and natural gas prices charged by several electric power utility companies and gas distributors in Brazil at the time this study was being carried out (November, 2002). Table 9 shows the general conditions used in the cases studies.

Table 9: General conditions of the analysis

Currency rate	2.6	R$/US$
Interest rate	10	% per year
CAPSTONE Microturbine Features		
Model	330 (High Pressure)	
Fuel	Natural Gas	
Proprieties of Natural Gas (20 °C and 1 atm)		
Specific Mass	0.602	
High Heat Value	39,304	kJ/^{m}3

Peak Shaving Case

Many consumers try to reduce their electricity consumption at peak hours due to its high price. If they can produce their electricity, they will reduce the amount of electricity purchased from utility companies at peak hours, without having to reduce their electricity consumption. Besides, power generation systems can improve the quality and reliability of the energy supplied by utility companies.

A study was carried out in four Brazilian regions, classified according to the price of natural gas charged by gas distributors of these regions, as shown in Table 10. Table 11 and Figure 18 show the conditions studied and the electricity demand supplied by utility companies with and without peak shaving.

Table 10: Brazilian regions analyzed in the peak shaving case

	Brazilian States
1st Region	São Paulo (SP) and Rio de Janeiro (RJ)
2nd Region	Ceará (CE), Pernambuco (PE) and Paraíba (PB)
3rd Region	Rio Grande do Norte (RN)
4th Region	Others

Commercial microturbines available in the Brazilian market are imported from the USA and investments feasibility depends on the currency rate, as can be seen in Table 9.

Table 11: Conditions of the peak shaving case

Model of microturbine	**Capstone 33**0	
Number of microturbines	1	
Life time of microturbines	20	years
Net power (peak load)	28	kW
Microturbine installed cost	1.538	US$/kW
Natural gas consumption (HHV)	650	m³/month
Average price of natural gas (taxes included)	0.33-1.32	R$/ᵐ3

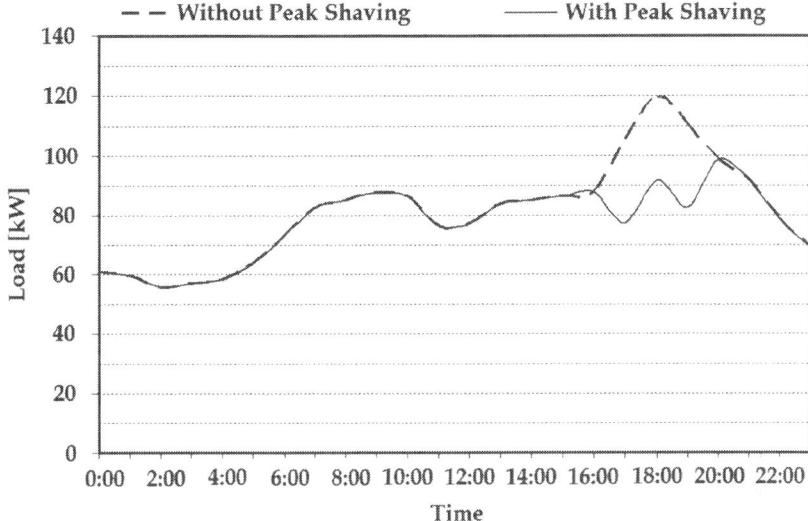

Figure 18: Electricity demand supplied by utility companies.

Table 12 displays the economical analysis of the peak shaving case. The investment is not feasible yet, as the payback period is very long. Rio Grande do Norte is the state where this business would be most interesting as payback is 8 years.

Table 12: Economical analysis of the peak shaving case

		SP - RJ	CE-PE-PB	RN	Other States
Total Investment *	US$	46,827	46,827	46,827	46,827
Annual cost**	US$/year	53,827	44,906	55,718	51,102
Annual cost*	US$/year	55,323	45,231	53,950	51,479
Annual savings	US$/year	-1,497	-325	1,769	-378
Electricity generated	US$/MWh	435	321	301	366
Payback Period	years	32	15	8	15

[i] - * With peak shaving; ** Without peak shaving

Figure 19 shows payback period in relation to microturbine cost. There is a strong fall on the payback period of the states of SP and RJ, due to a decrease in the microturbine cost.

A few manufactures intend to decrease microturbine costs to about 400 US$/kW until 2005 (Dunn & Flavin, 2000). If the microturbine cost is 400 US$/kW, the payback period will be between 2.5 and 5 years, as shown in Figure 19.

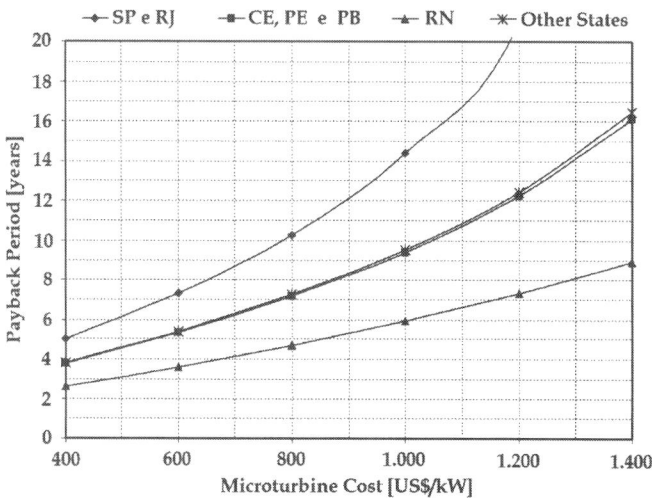

Figure 19: The influence of the microturbine cost on the return on investments.

Base Load Case

In this case, a microturbine produces electricity to a gas station according to the base load demand, as shows Figure 20. The conditions of this case are in table 13, whereas Table 14 shows the Brazilian regions analyzed in the base load case.

Table 13: Conditions of the base load case.

Model of microturbine	Capstone 330	
Number of microturbines	1	
Life time of microturbines	10	years
Net power	27,5	kW
Microturbine installed cost	1,538	US$/kW

| Natural gas consumption (HHV) | 6,918 | m³/month |
| Average price of natural gas (taxes included) | 0.24 - 1.02 | R$/m3 |

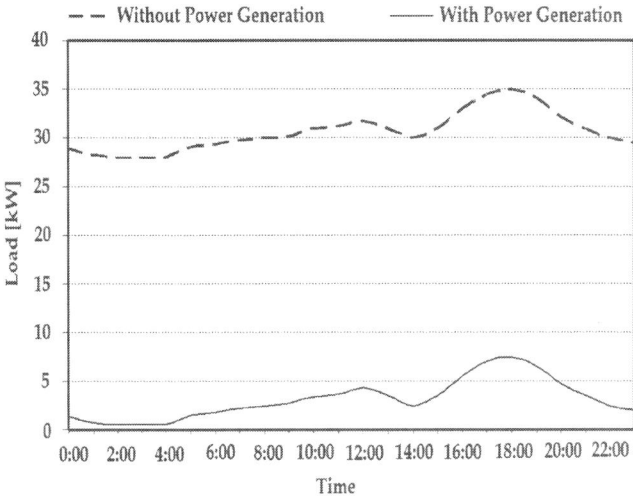

Figure 20: Electricity demand supplied by utility companies.

Table 14: Brazilian regions analyzed in the base load case

	Brazilian States
1st Region	São Paulo (SP) and Rio de Janeiro (RJ)
2nd Region	Rio Grande do Sul (RS) and Paraná (PR)
3rd Region	Rio Grande do Norte (RN)
4th Region	Others

Table 15 displays the economical analysis of the base load case for gas stations. Up to the present moment this kind of business is not feasible, except in the state of Rio Grande do Norte (RN) where payback period can be 3.1 years, once local gas distribution companies have encouraged thermoelectric small scale power generation, according to natural gas price lower than others kind of fuels.

Table 15: Economical analysis of the base load case

		SP e RJ	RS e PR	RN	Other States
Total Investment	US$	46827	46827	46827	46827
Annual cost**	US$/year	28748	28117	27956	24389
Annual cost*	US$/year	45699	33898	20707	26002
Annual savings	US$/year	- 16951	- 5780	7249	- 1614
Electricity generated	US$/MWh	181	131	75	99
Payback Period	years	Not Feasible	Not Feasible	3,1	8,2

[i] - * With power generation; ** Without power generation

Figure 21 shows the behavior of the cost of the electricity generated for different microturbine costs and natural gas average price. Some natural gas distribution companies in Brazil have encouraged the creation of small thermal power generation units, as the cost of natural gas coming from these companies would be about 0.24 R$/m³. Based on this fact and on the perspective of microturbine manufactures, Figure 21 shows the cost of the electricity generated could be 58 US$/MWh. For each 1 US$/kW decreased from the microturbine cost, the cost of the electricity generated decreases about 0,021 US$/MWh, for every natural gas average price range, and for each 1 R$/m³ decreased from the natural gas average price, the cost of the electricity generated decreases about 135 US$/MWh, for every microturbine cost range.

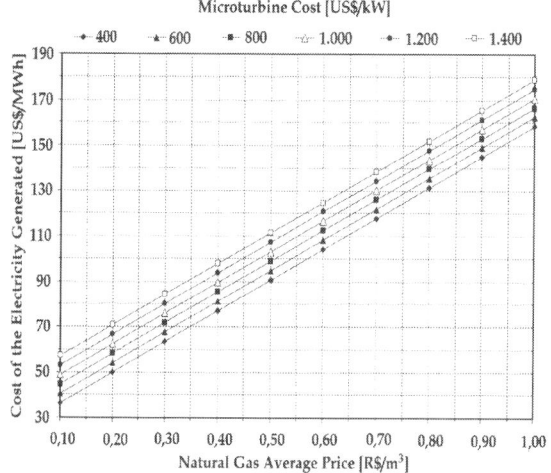

Figure 21: Cost of the electricity generated for different microturbine costs and natural gas average prices.

In the base load case, the natural gas average price is the most influential component in the return on investments. Figure 22 shows this conclusion for the microturbine cost at this moment, since natural gas average price of 0.24 R$/m^3 can result in a payback period between 3 and 4 years.

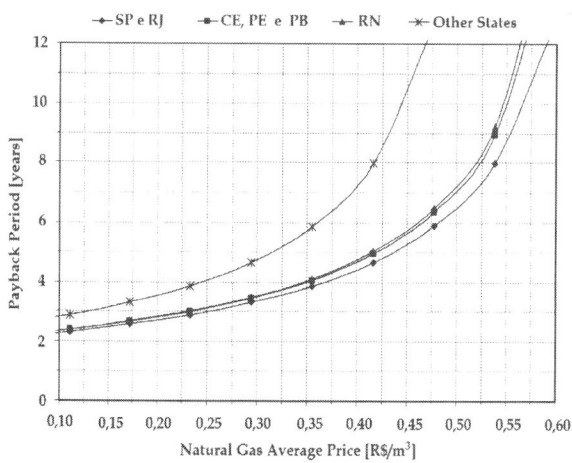

Figure 22: The natural gas average price influence on the payback period.

Cogeneration Case

In this case, two microturbines and a heat recovery system produced electricity and hot water to buildings in a residential segment, according to the base load demand, as can be seen in Figure 23.

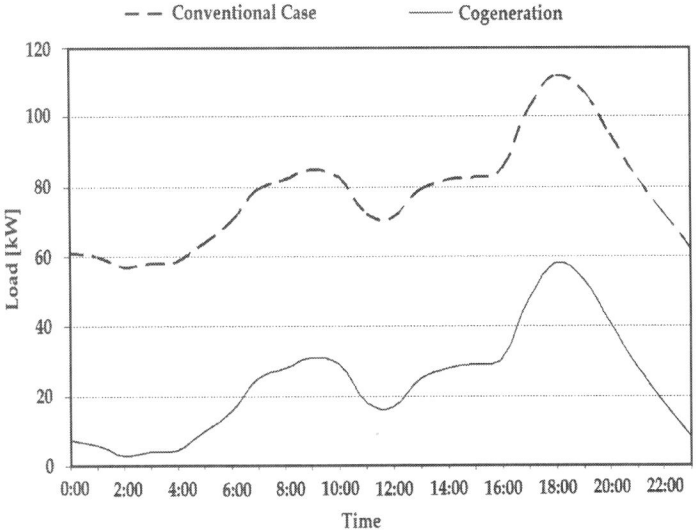

Figure 23: Electric demand supplied by utility companies to consumers with and without cogeneration.

A cogeneration plant can result in substantial savings of energy. However, these systems usually result in greater capital expenditures than non-cogeneration plants. This incremental capital investment for cogeneration must be justified by reduced annual energy costs and reduced payback periods.

A course of action involving minimum capital expenditures can be determined as the conventional case. In this study a low pressure boiler supplying process heat and the purchase of all electric power from utility system is the conventional case. Although the conventional case has the lowest investment cost, it usually has annual operating costs significantly higher than those available with cogeneration alternatives. Table 16 shows the conditions of this case, while Table 17 shows the Brazilian regions analyzed in the base load case.

Table 16: Conditions of the cogeneration case

System cogeneration model	MG2-C1	
Number of Capstone microturbines	2	
Number of heat recovery systems	1	
Life time of microturbines	10	years
Power output	54	kW
Heat recovery systems (hot water generation)		
Water pressure	10	bar
Water flow	2.22	t/h
Inlet water temperature	25	°C
Outlet water temperature	67	°C
Outlet exit gas temperature	93	°C
Net power	53	kW
System cogeneration installed cost	1,872	US$/kW
Natural gas consumption (HHV)	13,653	m^3/day
Average price of natural gas (taxes included)	0.24 - 0.90	R$/^{m}3

Table 17: Brazilian regions analyzed in the cogeneration case

	Brazilian States
1st Region	Rio de Janeiro (RJ)
2nd Region	Paraná (PR)
3rd Region	Rio Grande do Norte (RN)
4th Region	Others

Table 18 displays the economical analysis of the cogeneration case. Investments costs are lower in the conventional case than in the cogeneration system, and, although the annual cost is higher, savings can be up to US$ 24,907 per year. The payback period is between 2.8 and 3.8 years and the minimal cost of the electricity generated is 84 US$/MWh.

Table 18: Economical analysis of the cogeneration case

		RJ	PR	RN	Other States
Total Investment*	US$	23077	23077	23077	23077
Total Investment**	US$	136797	136797	136797	136797
Annual cost*	US$/year	128566	110022	90323	98328
Annual cost**	US$/year	110337	96655	65416	77534
Annual savings	US$/year	18228	13367	24907	20795
Electricity generated	US$ / MWh	174	146	84	112
Payback Period	years	3,3	3,8	2,8	3,1

[i] - * Conventional; ** Cogeneration

In the cogeneration case, the fuel cost is the most influential component on the return on investment, similar to the base load case. Figure 24 shows fuel costs can represent up to 71% of the cost of the electricity generated.

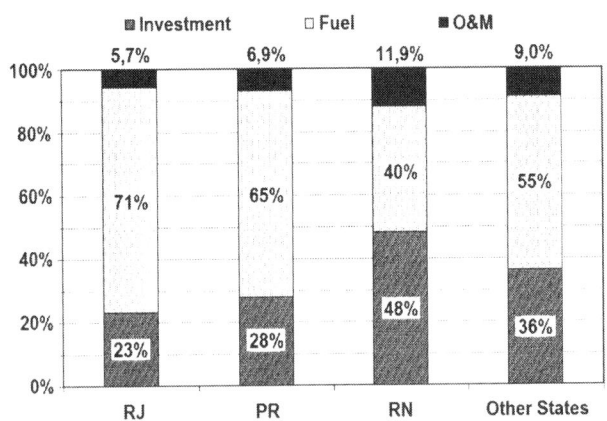

Figure 24: Components of the cost of the electricity generated.

Figure 25, Figure 26 and Figure 27 show the combined influence of microturbine cost and the average price of natural gas on the return on investment in the states of Rio de Janeiro and Paraná (Figure 25), Rio Grande do Norte (Figure 26) and the other states (Figure 27). Based on the perspective of microturbine manufactures and with natural gas average price of 0.25 R$/m³, the payback period can be between 1.5 and 3 years.

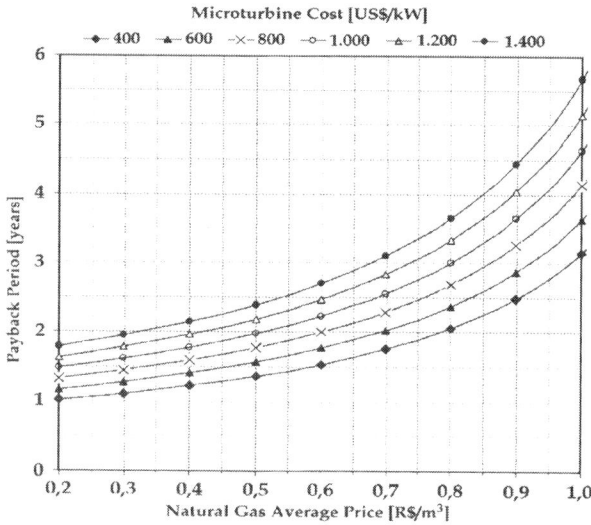

Figure 25: Combined influence of microturbine cost and average price of natural gas on the payback period in the states of Rio de Janeiro and Paraná.

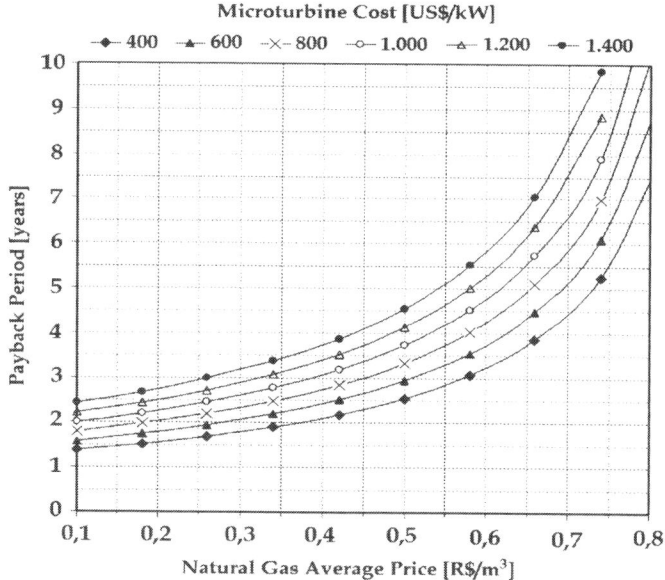

Figure 26: Combined influence of microturbine cost and average price of natural gas on the payback period in the state of Rio Grande do Norte.

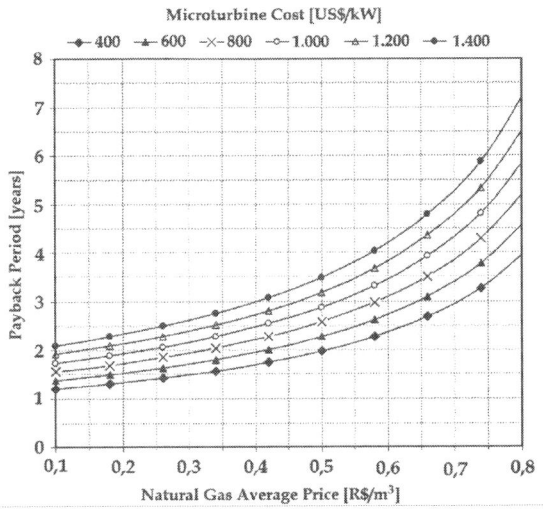

Figure 27: Combined influence of microturbine cost and average price of natural gas on the payback period in the other states.

CONCLUSIONS

The variable speed operation and the electric power conditioner increase part-load efficiency of microturbines as they allow for the improvement of part-load fuel savings, especially increased recuperator effectiveness at lower part-load airflows. The variable speed control improves part-load performance but requires a system able to sense load and optimize speed. According to the results shown in this study, the microturbines efficiency is at its highest when Capstone microturbines are operating over an output range between 12 kW and full load.

Capstone microturbines use clean combustion technology to achieve low emissions. Nitrogen oxides (NO_x) and carbon monoxide (CO) emission levels of these machines are lower than 7 ppmv@15%O_2 at full load when these microturbines are fueled with natural gas.

Microturbines exhibit low emissions of all classes of pollutants and have environmental benefits as they release fewer emissions compared to other distributed generation technologies, like internal combustion engines. Besides, these units are clean enough to be placed in a community with residential and commercial buildings.

Microturbine generators have shown good perspectives for electricity distributed generation in small scales, once they have high reliability and simple design (high potential for large scale cheap manufacturing).

Although results show microturbines are not feasible to provide energy at peak demand, in this case the microturbines can supply peak demand and improve the level of reliability of the electricity supplying, because they can provide stand-by capabilities should the electric grid fail.

In the base load case this sort of business is feasible just in states of Brazil where natural gas distributing companies have encouraged small thermal power generation by natural gas with lower prices, since the price is the most influential cost component of the electricity generated.

The most feasible investment in microturbines is in the cogeneration case. In this case, economical feasibility is certain in all states of Brazil as cogeneration systems can provide considerable annual savings. Besides, under the perspective of manufacturers, and with the incentive of natural gas distribution companies together with the rise in electricity prices of Brazilian utility companies, investments in microturbines for the next years will be higher than currently.

ACKNOWLEDGEMENTS

The authors would like to thank CAPES, FAPEMIG, FAPEPE and CNPq, for their financial support.

REFERENCES

1.   Asme performance test code PTC-22-1997, Gas turbine power plants, 1997.

2.   Barker, T. Micros, Catalysts and Electronics, Power-Gen International 96, Turbomachinery, v. 38, n. 1, p. 19-21, 1997.

3.   Biasi, V. de Low cost and high efficiency make 30 to 80 kW microturbines attractive, Gas Turbine World, Jan.-Feb., Southport, 1998.

4.   Bolszo, C. D., and Mcdonell, V. G., Emissions Optmization of a Biodiesel Fired Gas Turbine, Proceedings of the Combustion Institute, 32, LSEVIER, pp. 2949-2956, 2009.

5.   Capstone Turbine Corporation, Capstone Low Emissions Microturbine Tecnology, White Paper, USA, 2000.

6.   Capstone Turbine Corporation, Capstone Microturbine Model 330 System Operation Manual, USA, 2001.

7.   Capstone Turbine Corporation, Capstone Microturbine Product Catalog, USA, 2012: http://www.capstoneturbine.com/prodsol/products/, accessed at: 20/06/2012.

8.   Cohen, H., Rogers, G. F. C., and Saravanamuttoo, H. I. H., Gas Turbine Theory, Fourth edition, 1996.

9.   Dunn, S. & Flavin, C., Dimensionando a Microenergia. In: Estado do Mundo 2000. Brazil, UMA Ed., 2000.

10. Gomes, E. E. B. Análise Técnico-econômica e Experimental de Microturbinas a Gás Operando com Gás Natural e Óleo Diesel, Master Degree Thesis, Supervised by Nascimento, M. A. R. and Lora, E. E. S. Federal University of Itajubá, 2002.

11. Hamilton, S. L., Microturbines, Distributed Generation: a nontechnical guide, edited by Ann Chambers, cap. 3, pp. 33 – 72, PennWell Corporation, USA, 2001

12. GRI - Gas Research Institute, The role of Distributed Generation in competitive energy markets, Distributed Generation Forum, Gas Research Institute (GRI), 1999.

13. Liss, W.E., Natural Gas Power Systems for the Distributed Generation Market. Power-Gen International '99 Conference. CD-Rom. New Orleans, Louisiana, USA, 1999.

14. Nascimento, M. A. R.; Santos, E. C., Biofuel and Gas Turbine Engines, Advances in Gas Turbine Technology, InTech, ISBN 978-953-307-611-9, chaper 6, 2011.

15. Nwafor, O., Emission characteristics of Diesel engine operating on rapeseed methyl ester. Renewable Energy, 29, pp. 119-29, 2004.

16. Pierce, J. L. Microturbine Distributed Generation Using Conventional and Waste Fuel, Cogeneration and On-Site Power Production, James & James Science Publishers, p. 45, v. 3, Issue 1, Jan-Feb, 2002.

17. Rodgers, C.; Watts, J.; Thoren, D.; Nichols, K. & Brent, R. Microturbines, Distributed Generation – The Power Paradigm for the New Millennium, edited by Anne-Marie Borbely & Jan F. Kreider, cap. 5, pp. 120 – 148, CRC Press LLC. USA, 2001a.

18. Rodgers, G., and Saravanamutto, H., Gas Turbine Theory, Prentice Hall, 2001b.

19. Scott, W. G. Micro Gas Turbine Cogeneration Applications, International Power and Light Co., USA, 2000.

20. Sierra, R. G. A., Teste Experimental e Análise Técnico-Econômica do Uso de Biocombustíveis em uma Microturbina a Gás de Tipo Regenerativo; Dissertação de Mestrado, UNIFEI, 2008.

21. Watts, J. H, Microturbines: a new class of gas turbine engine, Global Gas turbine News, ASME-IGTI, v. 39, n. 1, p. 4-8, USA, 1999.

22. Weston, F., Seidman, N., L., James, C. Model Regulations for the Output of Specified Air Emissions from Smaller-Scale Electric Generation Resources, The Regulatory Assistance Project, 2001.

Vibration Cause Analysis and Elimination of Reciprocating Compressor Inlet Pipelines

Zheng Liang[a], Shuangshuang Li[a], Jialin Tian[a], Liang Zhang[a], Chengke Feng[b], and Liwen Zhang[c]

[a]School of Mechatronic Engineering, Southwest Petroleum University, Chengdu, Sichuan 610500, China
[b]Chongqing Division of Southwest Oil & Gasfield Company, PetroChina, Chongqing 400021, China
[c]Baoji Oilfield Machinery Co., Ltd, PetroChina, Baoji, Shaanxi 721002, China

ABSTRACT

Due to the occurrence of abnormal vibration of reciprocating compressor inlet pipelines during the commissioning of a booster station, the cause of severe vibration problem was investigated, which included modal analysis, calculation of resonant piping length, velocity frequency spectrum analysis, and pressure pulsation measurement. It was found that the inlet pipelines avoided low frequency resonance region, the actual length of the inlet pipelines was in the second resonant piping length, and the pressure pulsation far exceeded API 618 standard. The results indicate that large pressure pulsation and acoustic resonance occurred on the inlet pipelines are the key factors inducing vibration. Vibration elimination treatments included enlarging the buffer volume of gathering manifold, adjusting the inlet piping length to avoid acoustic resonance, and increasing the curvature radius of bend. After remodeling of the inlet pipelines, the test data indicate that the vibration level of the inlet pipelines is reduced to an acceptable level defined by a relevant standard, and the processing capacity of the booster station can be raised at least two times.

INTRODUCTION

Reciprocating compressors are widely utilized in the natural gas transmission and natural gas industries because they are flexible in throughput capacity and discharge pressure range. However, large gas fluctuations can be induced because of the intermittent suction/discharge flow. The resulting pressure pulsations in piping system can induce severe piping vibration at a discontinuous region such as elbow, reducer, tee branch or valve. Abnormal or excessive vibrations lead to fatigue failure of piping system, the overloading of the compressor and other safety problems. For example, at a booster station, its design processing capacity was 183×10^4 m^3/d, and there were four reciprocating compressors, the power of compressor Nos. 1–3 were 1250 kW, and compressor No.4 was 1030 kW. Due to unreasonable design, the inlet pipelines, as shown in Fig. 1, generated severe vibration. The maximum vibration velocity was 34.26 mm/s, which far exceeded the standard API 618 that is 17.8 mm/s. Worse still, only one compressor could run because of the resulting severe vibration. In order to ensure production, one measure pouring cement on the pipe was taken to control vibration without any rational diagnosis and analysis, as illustrated in Fig. 2. Unfortunately, it could not be controlled radically, there were still severe vibrations. As a result, reducing and controlling vibration level of the piping become significant in engineering.

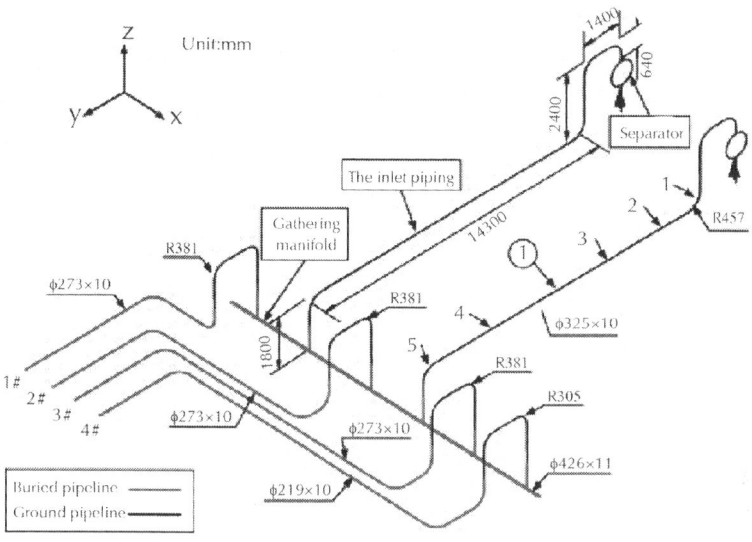

Figure 1: Layout of the inlet pipelines.

Figure 2: Field picture of the original inlet pipelines after pouring cement.

The studies concerning pipeline vibration and pressure pulsation mainly include theoretical research and engineering application. Theoretical research includes building mathematical modeling of pipelines, model parameters calculation, pressure pulsation simulation, fluid–structure interaction conveying fluid and a dynamic response of pipeline [1] and [2]. The acoustic wave theory, transfer matrix method, and finite element method have been proposed to analyze gas pulsation in the piping system [3] and [4]. Engineering application mainly focuses on vibration testing technique, vibration analysis technique, fault diagnosis technology, piping design and elimination solution [5] and [6]. To put an end to the vibration problems, this paper will do model analysis, site measurements and spectrum analysis to discern the key causes of pipe vibration. Meanwhile, according to the key causes and site conditions of the inlet pipelines, some practical elimination measures are going to be taken, and the effect of vibration elimination is to be evaluated after remodeling.

SITE MEASUREMENTS AND CAUSE ANALYSIS

There are two vibration systems in a compressor piping system: one is a mechanical structure system, and the other is an acoustic system. Each elastic system has mechanical structure frequencies and acoustic natural frequencies. When it is coincident with an excitation frequency or pulsation frequency of compressor, and the resulting resonance can induce severe piping vibration. Typical sources of unwanted vibration of piping systems include: (1) pressure pulsations and fluctuations, (2) structural vibration, (3) dynamic effect caused by rotating parts, (4) unbalance, misalignment, friction, impacts, etc. Therefore, it is necessary to identify possible vibration sources and find the key causes inducing piping vibration.

Mechanical Resonance Analysis

The study of mechanical structure resonance mainly concerns the relationship between natural frequencies of piping systems and excitation frequency of compressors. Modal analysis is an effective method to investigate natural frequencies and mode shapes of vibration [7]. As the rotational speed and excitation frequency of the reciprocating compressor are low in many cases, and the piping vibration is low frequency vibration, the evaluation of frequencies corresponding to first six modes can meet the engineering requirements [8]. There are two methods to calculate modal parameters (natural frequencies and mode shapes): one is finite element method, the other is an experimental method.

Finite Element Modal Analysis

In Fig. 1, the original inlet pipelines contain ground pipelines and buried pipelines. The ground pipelines are comprised of almost two symmetrical pipelines, one is in a closed state, and the other is in a working state in actual work. The finite element modal analysis of the ground pipelines was discussed. To minimize the computational consumption, 1/2 three-dimensional model of the original ground pipelines was established based on the actual structural parameter and position of the supports. The key parameters of the original inlet pipelines are as shown in Fig. 1. The grid type was hexahedral element and element type was "hex dominant". The boundary condition of buried pipelines was elastic support or assumed to be zero as the initial condition. The connection between the pipelines can be considered as bonded or frictional connection. The natural frequencies of the original model were obtained under the actual condition. Table 1 shows the first six natural frequencies using finite element method. Fig. 3 presents the first six mode shapes.

Table 1: The first six natural frequencies of the original model using finite element method

No.	1	2	3	4	5	6
Frequency (Hz)	2.8727	3.3024	5.4101	7.1316	8.2758	12.199

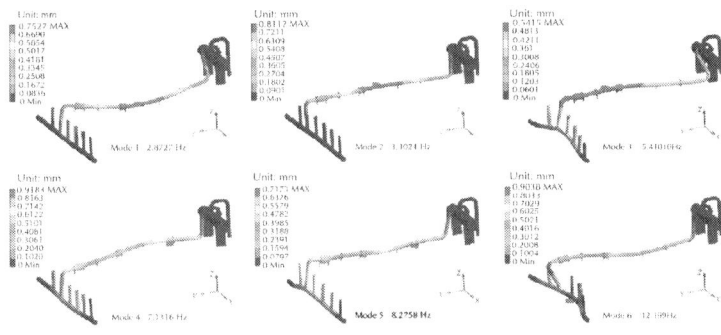

Figure 3: The first six mode shapes of the original model.

Experimental Modal Analysis

Experimental modal test was conducted to verify the results of finite element method. The natural frequencies of the original inlet pipelines were gained by the hammering method when the compressor did not run [9]. The locations of excitation points and collecting points are where large pipe vibration is caused, as shown in Fig. 4. The excitation points were excited with hammer in the x, y, and z direction respectively. Simultaneously, the acceleration signals from collecting points in each direction were collected by acceleration sensors. Every excitation point was necessary to be excited more than three times in each direction, and avoid forming a "regular rhythm". The first six natural frequencies of the original inlet pipelines were obtained by analyzing the frequency spectrum of the acceleration signals, as listed in Table 2.

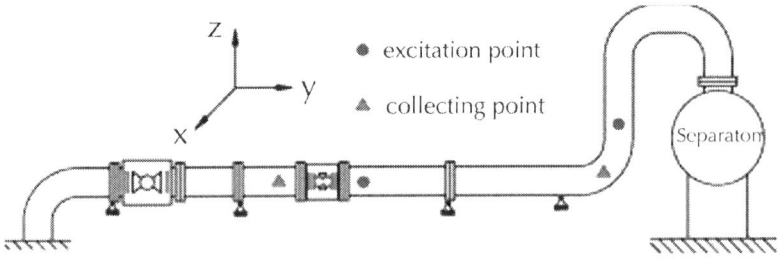

Figure 4: The locations of excitation points and collecting points.

Table 2: The first six natural frequencies of the original inlet pipelines using experimental method

No.	1	2	3	4	5	6
Frequency (Hz)	1.875	2.5	4.375	6.875	8.125	10.625

It is well-known that the resonances are most likely to occur when a particular natural frequency matches the excitation frequencies of compressor, thus, vibrations are strongly amplified. The excitation frequency f_{ex} is generally related to the rotational speed N and can be calculated by Eq. (1) [10]. The compressor was a single acting compressor cylinder and its rotational speed range was 950–1050 rpm, so the range of excitation frequency was 15.83–17.5 Hz.

$$f_{ex} = \frac{N}{60} ki$$

(1)

From Fig. 5, the first six natural frequencies of the original inlet pipelines using finite element method show good agreement with the experimental results, which indicates that the proper 3D model was established. However, some deviations remain, and for the following reasons: (1) the restraints in the 3D model are completely rigid, which in an actual piping system are elastic, therefore it cannot accurately simulate geometry features and constraints of an actual piping system, which lead to some deviation from the exact results. (2) Another reason for deviation is the measurement accuracy. Some natural frequencies appeared within a very close frequency range where measurement may not able to capture the peak value of frequency response.

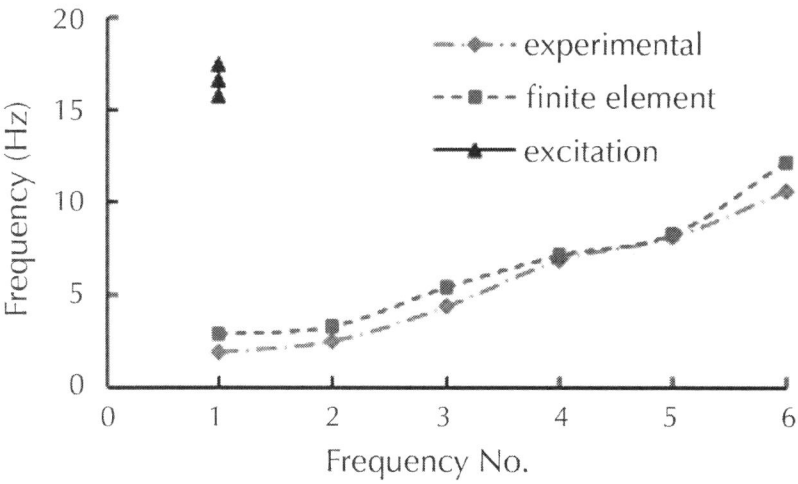

Figure 5: Frequencies comparison of the inlet pipelines.

Due to the first six natural frequencies of the original piping system being below the compressor excitation frequency avoided low frequency resonance region, it can be concluded that the serve piping vibration is not induced by mechanical structure resonance.

Acoustic Resonance Analysis

The gas in pipe has a certain acoustic natural frequency, when it matches the excitation frequency of the compressor, acoustic resonance occurs and oscillations are strongly amplified. The transfer matrix method is a useful method of calculating acoustic natural frequency in piping systems [11] and [12]. In this method, the piping system is divided into several piping elements such as a pipe, a side branch, and a volume. Each piping element has a transfer matrix M. For example, to a straight pipe, the transfer matrix M is given by Eq.(2). For the total piping system, the governing equation of acoustic natural frequency of pipe is formulated by multiplying the transfer matrices of all the piping elements as Eq. (3) [13].

$$M = \begin{bmatrix} \cos(\omega L/a) & -\rho a \sin(\omega L/a) \\ \frac{1}{\rho a} \sin(\omega L/a) & \cos(\omega L/a) \end{bmatrix} \qquad (2)$$

$$\begin{bmatrix} P_{end} \\ u_{end} \end{bmatrix} = M_n \cdot M_{n-1} \cdots M_2 \cdot M_1 \cdot \begin{bmatrix} P_{strt} \\ u_{strt} \end{bmatrix} \qquad (3)$$

The subscripts "strt" and "end" represent boundary conditions at the "starting point" and "end point" of a piping system respectively. The acoustic natural frequency is calculated by solving the frequency equations, which are obtained by giving boundary conditions (open or closed) at the starting point and the end point of the piping system.

Pipeline connected to small volume can be regarded as a closed end, in which the boundary conditions are the velocity fluctuation is zero ($u = 0$) and the pressure fluctuation is constant ($p = C$). Pipeline connected to large volume can be regarded as an open end, in which the boundary conditions are the pressure fluctuation is zero ($p = 0$) and the velocity fluctuation is constant ($u = C$) [14]. Using these boundary conditions, the acoustic resonance frequency of a straight pipe can be estimated from Eq. (4) [15].

$$\begin{cases} f_0 = (a/2L)i & \text{(open–open or closed–closed)} \\ f_0 = (a/4L)(2i-1) & \text{(open–closed)} \end{cases} \qquad (4)$$

The sound speed of real gas can be obtained by Eq. (5)[16].

$$a = \sqrt{k_v Z R_g T} \qquad (5)$$

The resonant piping length "L" refers to the length of pipe when the acoustic resonance occurs, and can be calculated by rearranging Eqs. (4) and (5) as shown in Eq. (6). All multiples of "L" will develop resonant frequencies. As the first acoustic frequency is the strongest and most destructive, piping length should be much shorter or longer than a multiple of the resonant piping length [17].

$$\begin{cases} L = (\sqrt{k_v Z R_g T}/2f_{ex})i & \text{open–open or clos} \\ L = (\sqrt{k_v Z R_g T}/4f_{av})(2i-1) & \text{(open–closed)} \end{cases} \qquad (6)$$

where $L = L_f + L_c$, L_f is fundamental length of straight piping, L_c is the conversion length of bend, tee branch, etc.

From Fig. 1, the boundary condition of the inlet pipelines is open-closed. From Eq. (6), the resonant piping length "L" is relate to the parameters of real gas in pipe. The parameters of the feed gas of compressor are given in Table 3. As a result, the first six resonant piping length listed in Table 4 is obtained by substituting the data in Table 3 into the Eq. (6). However, the actual length of the inlet pipe was 18.51 m according to site investigation, which was in the second resonant piping length based on the data in Table 4, which may lead to acoustic resonance and induce pipeline vibration.

Table 3: The parameters of the feed gas of compressor

k_v	Z	R_g (J/(kg K))	T (K)	a (m/s)	f_{ex} (Hz)
1.29	0.9192	487.1	298.15	414.97	15.83–17.5

Table 4: The first six resonant piping length "L" of the original inlet pipelines

No.	1	2	3	4	5	6
L (m)	5.928–6.553	17.784–19.661	29.641–32.768	41.497–45.875	53.354–58.983	65.211–72.091

Site Vibration Measurements

In order to further determine the cause and severity of vibration, site vibration measurements were made on the inlet pipelines according to standard ISO 10816-6 [18]. Vibration measurements include amplitude, velocity, acceleration and pressure pulsation. Fig. 1 shows the vibration measuring points 1–5 on the inlet pipelines. To test real pipeline vibration data, the compressor must work steady at least 15 min before measured.

In the test cases, the rotational speed of compressor No. 4 was set at 950 rpm, 1000 rpm and 1050 rpm respectively, under two different operating conditions (idle load and load). Speed fluctuation is inevitable as the compressor is driven by gas engine. Table 5 is the operating parameters of compressor No. 4. Vibration test results with the rotational speed of compressor No. 4 compressor at 1047 rpm are seen in Table 6. A vibration velocity comparison under two different operating conditions is given in Fig. 6. The velocity frequency spectrum of measuring point 2 is given in Fig. 7, which is similar to the results of other points.

Table 5: The operating parameters of compressor No. 4

Speed (rpm)	Inlet pressure (MPa)	Discharge pressure (MPa) =		Discharge temperature (°C) =		Processing capacity (m³/d)
		First stage	Second stage	First stage	Second stage	
1047	4.28	6.83	6.85	70.5	78.6	646,138

Table 6: Vibration test results with the rotational speed of compressor No. 4 at speed of 1047 rpm

Operating case	Measuring point	Amplitude (µm)			Velocity (mm/s)			Acceleration (m/s²)		
		x	y	z	x	y	z	x	y	z
Idle load	1	44.33	11.23	24.76	4.25	1.23	2.6	0.535	0.138	0.288
	2	42.9	10.79	10.26	3.95	1.12	1.06	0.512	0.125	0.116
	3	39.95	13.02	5.9	3.83	1.36	0.64	0.477	0.152	0.072
	4	21.05	12.56	6.79	2.1	0.84	0.51	0.230	0.092	0.060
	5	13.21	8.12	6.72	1.42	0.89	0.23	0.156	0.099	0.038
Load	1	14.07	72.4	61.63	1.67	8.09	4.85	0.330	1.31	1.580
	2	39.67	101.53	18.67	2.04	11.15	0.68	0.346	1.231	0.210
	3	20.47	101.99	2.25	2.36	11.2	0.26	0.365	1.236	0.120
	4	50.55	104.32	14.3	0.69	11.45	0.93	0.083	1.261	0.165
	5	24.96	48.46	68.35	2.7	5.23	4.51	0.370	0.586	0.840

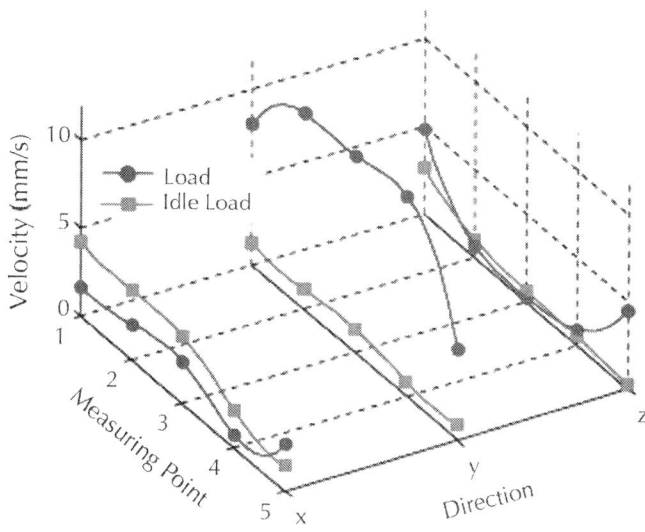

Figure 6: Vibration velocity comparison under two different operating conditions

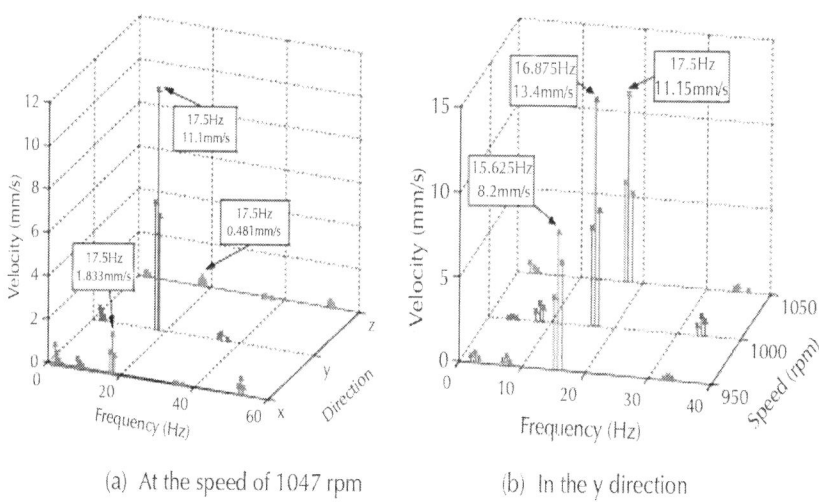

(a) At the speed of 1047 rpm   (b) In the y direction

Figure 7: Velocity frequency spectrum of measuring point 2.

A simple way to know whether vibration is caused by pressure pulsation is to compare vibration severity under the two different working conditions (idle load, load). The vibration is mainly caused by pressure pulsation if vibration severity is weak under idle load condition, but is strong under load condition. FromTable 6 and Fig. 6, the maximum vibration velocity under load case increases by 270% than that under idle load case, especially in the y direction. Therefore, severe pipeline vibration is caused by pressure pulsation in a preliminary estimate.

The purpose of spectrum analysis is to get the dominant frequency components. Additionally, frequency spectrum plot is a "must have" in determining the severity of the vibration and the sources of vibration. The vibration velocity frequency spectrum contain many harmonic components as shown in Fig. 7, and the velocity frequency peak in the spectrum is very close to the excitation frequency of compressor, however, there are weak vibrations in other harmonic components. These spectrum characteristics are closely related to pressure pulsations.

The pressure pulsation can be evaluated by pressure unevenness. API618 dictates the maximum allowable level of pressure pulsation, which can be calculated by Eq. (7)[19]. The location of pressure pulsation point ① is presented in Fig. 1. The test results show that the maximum pressure unevenness of point ① is 2.04%. However, the maximum allowable level of pressure pulsation is 0.91% by Eq. (7), so this pressure unevenness far exceeds the standard.

$$P_1 = \sqrt{\frac{a}{350}} \frac{126.5}{\sqrt{P_L \cdot ID \cdot f_{ex}}} \tag{7}$$

In summary, the root cause of the abnormal pipeline vibration is excessive pressure pulsation and the inlet pipelines length in the second resonant piping length. It must be mentioned that several major defects in the original inlet pipelines were found in site observation: (1) the diameter of the gathering manifold was too small and its buffer volume was limited; moreover, there was no drain outlet. With the accumulation of dirt and impurities, the effective buffer volume of gathering manifold became smaller and smaller; (2) the design of the inlet and outlet of gathering manifold and the location of the pipe clamp on the inlet pipes were not reasonable.

VIBRATION ELIMINATION TREATMENTS

To gas pipeline systems, neglecting the interaction between pipe wall and gas, pipe vibration induced by periodic excitation force acting on pipe can be considered as forced vibration. The governing equation of motion of pipeline system is given by Eq. (8).

$$[M]\{\ddot{x}\} + [C]\{\dot{x}\} + [K]\{x\} = \{F\}$$

(8)

From Eq. (8), there are some approaches to eliminating or attenuating vibration: (1) changing structure characteristic parameters [M], [C] and [K], accomplished by using the vibration isolator and increasing the support stiffness; (2) reducing excitation force {F}, such as increasing curvature radius of elbow, enlarging buffer volume, and using an orifice; (3) adjusting mechanical frequency of the pipe system to avoid low frequency resonance, and isolating the resonant piping length or modifying the boundary condition (open or closed). Vibration induced by pressure pulsation is not easy to be controlled, because it cannot eliminate vibration force caused by pressure fluctuation through increasing support stiffness or adding support. Therefore, according to the processing requirement and actual field conditions, and considering the above analysis, the treatments are as follows:

using two independent gathering manifolds to meet the requirements of equipment maintenance; (2) the nominal diameter of gathering manifold is 800 mm (DN800) and its buffer volume is 2.56 m^3, which is based on the flow area of gathering manifolds is at least three times greater than all inlet pipes; (3) decreasing the use of bend and increasing curvature radius to R = 5DN; (4) adjusting the inlet pipelines length to 13.75 m and mounting supports reasonably. The layout of the new inlet pipelines for vibration elimination treatments is as shown in Fig. 8.

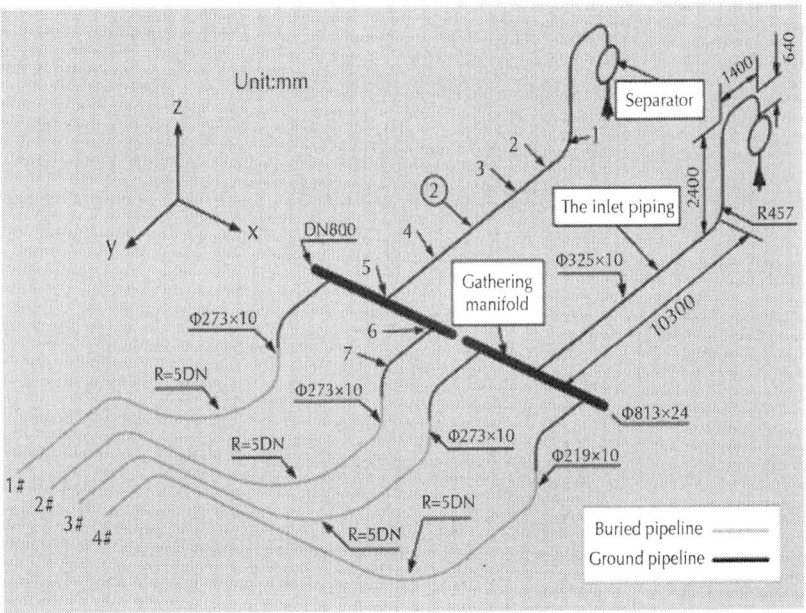

Figure 8: The layout of the new inlet pipelines for vibration elimination treatments.

As noted earlier, the new inlet pipelines also contain ground pipelines and buried pipelines as seen in Fig. 8. To estimate the vibration level of the new inlet pipes after remodeling, modal analysis and harmonic analysis were carried out on the 3D model of the ground pipelines. The key parameters of the new inlet pipelines are as shown in Fig. 8. The settings of element type and boundary conditions were the same as that for model analysis of the original inlet pipelines. The vibration velocity can be gained exactly under the results of modal analysis by harmonic analysis. Fig. 9 represents the first six natural frequencies and mode shapes. Fig. 10 is the results of harmonic analysis under 17.5 Hz.

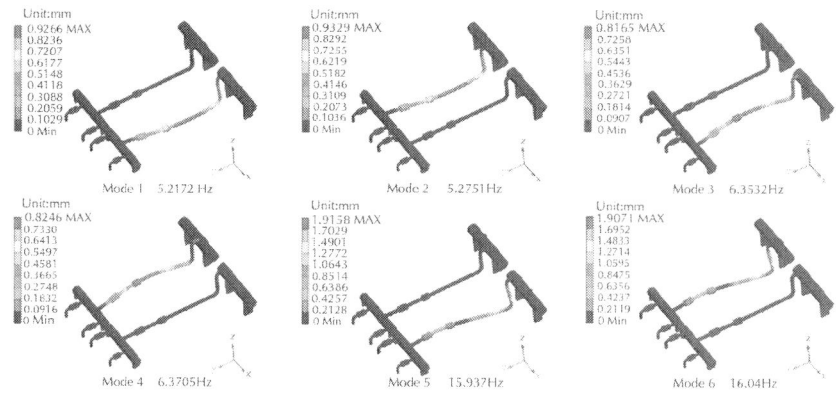

Figure 9: The first six natural frequencies and mode shapes of the new model.

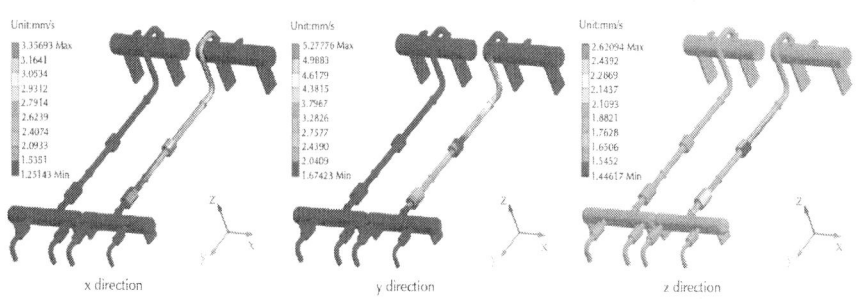

Figure 10: The results of harmonic analysis under 17.5 Hz of the new model.

As shown in Fig. 10, the maximum vibration velocity is 5.2776 mm/s in the y direction, which meets the API618 standard. The remodeling of the inlet pipelines was completed based on Fig. 8. Comparison of site photo before and after remodeling is seen in Fig. 11. As mentioned above, site vibration measurements were completed in two cases to evaluate the effect of vibration attenuation after remodeling. The first case saw the compressor No. 4 running alone, and the second case saw the compressor Nos. 1 and 2 running simultaneously. The operating parameters of compressors after remodeling are seen in Table 7. The location of vibration measuring points 1–7 and pressure pulsation measuring point ② after remodeling are shown in Fig. 8. The vibration measurement results under two cases after remodeling are listed in Table 8.

Figure 11: Comparison of site photos before and after remodeling.

Table 7: The operating parameters of compressors after remodeling

Case	Compressor No.	Speed (rpm)	Inlet pressure (MPa)	Discharge pressure (MPa)		Discharge temperature (°C)		Processing capacity (m³/d)
				First stage	Second stage	First stage	Second stage	
First	4#	1048	4.21	6.55	6.57	73.6	76.8	624,197
Second	1#	1046	4.13	5.06	6.40	41.0	63.8	1,195,221
	2#	1051	4.10	4.97	6.41	42.6	60.1	

Table 8: The vibration measurement results under two cases after remodeling

Case	Measuring point	Amplitude (μm)			Velocity (mm/s)			Acceleration (m/s²)		
		x	y	z	x	y	z	x	y	z
First	1	31.08	71.66	24.33	2.76	4.42	2.68	0.478	0.654	0.514
	2	40.71	34.72	20.06	2.37	3.86	2.17	0.491	0.506	0.432
	3	17.52	34.83	13.78	1.89	3.87	1.27	0.478	0.519	0.427
	4	92.96	28.35	14.49	1.79	3.17	1.47	0.362	0.621	0.385
	5	7.2	28.64	5.97	0.65	3.15	0.72	0.493	0.661	0.383
	6	25.45	18.93	90.76	0.88	1.88	2.34	0.232	0.413	0.547
	7	17.02	15.35	22.2	2.62	1.81	2.78	0.679	0.411	0.719
Second	1	46.93	33.61	20.28	5.52	3.91	2.41	0.89	0.66	0.55
	2	32.76	20.97	4.69	3.83	2.61	1.12	0.63	0.57	0.48
	3	12.65	21.96	2.4	1.35	2.63	0.58	0.36	0.57	0.32
	4	30.66	36.5	3.05	4.63	4.13	0.77	0.96	0.74	0.34
	5	47.16	35.44	5.93	5.22	3.97	0.95	0.74	0.57	0.5
	6	54.22	26.71	18.5	5.89	3.22	2.21	0.79	0.64	0.58
	7	58.8	37.67	42.23	6.54	4.2	5.11	1.36	0.68	1.09

As seen in Table 8, the vibration amplitudes are all reduced far below 283 μm, and vibration velocities are controlled below 17.8 mm/s, which meet the requirements of the standard ISO 10816-6 and API 618. FromFig. 12, the vibration velocities are significantly lower than those before remodeling in the y direction, and the largest decrease is 61.4%. Meanwhile, the maximum pressure unevenness of point ② is 0.64%, which is reduced within the maximum allowable level being 0.91%. The vibration severity of the inlet pipelines is much improved after remodeling, which can meet the requirements of running two or more compressors simultaneously, so the processing capacity of the booster station receives a considerable upgrade.

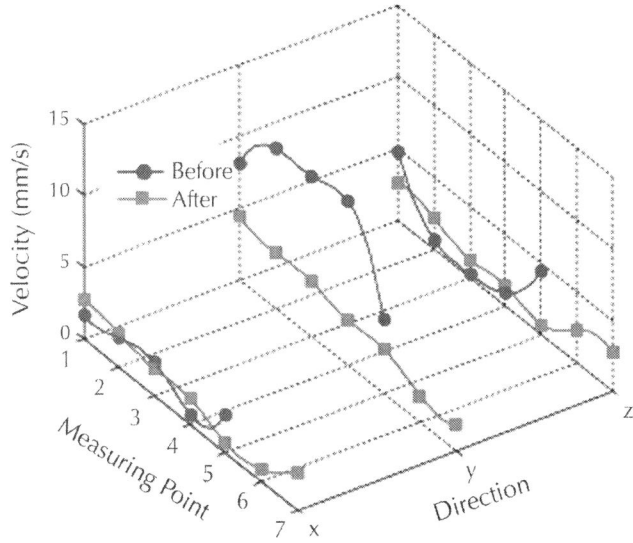

Figure 12: Comparision of vibration velocity before and after remodeling when compressor No. 4 ran alone.

CONCLUSIONS

The most important conclusions of the present paper are as follows.

- The vibration cause was investigated to solve the large inlet pipelines vibration problem. The first six natural frequencies of the inlet original pipelines were obtained by finite element modal analysis and experimental modal test using hammering method. The first six resonant piping length "L" was calculated by transfer matrix method. Site measurements during various working conditions of compressors were conducted to analyze velocity frequency spectrum, vibration severity, and pressure unevenness. The results show that it is not mechanical resonance, but the pressure pulsation and inlet pipelines length in the second resonant piping length that induce excessive pipe vibration.

- To control pressure pulsation and avoid acoustic resonance, some practical elimination treatments including enlarging the buffer volume of gathering manifold and adjusting the inlet piping length were implemented. The results of harmonic analysis and site vibration measurements show that the vibration amplitude, velocity, and pressure pulsation meet standard requirements after remodeling. It shows that the treatments adopted indeed eliminate vibration sources.

- In engineering application, at piping design stage, the layouts of pipe need to meet process requirements and vibration standards, so it is necessary to calculate the modal parameters and dynamic response of the piping system to optimize piping configuration; for existing pipelines, the key factors inducing excessive vibration can be found out easily by site measurement under different operating conditions of compressor and velocity frequency spectrum analysis, so the cost-effective solutions can be taken to reduce vibration. As pressure pulsation is inevitable owing to the intermittent suction/discharge flow of reciprocating compressor, the primary condition is to control gas pulsation and decrease excitation force acting on pipe.

ACKNOWLEDGEMENTS

This work is supported by the Innovation Funds for Postgraduate of School of Mechanical Engineering of Southwest Petroleum University (Grant No. CX2014BY04).

REFERENCES

1. Xu B, Feng Q, Yu X. Prediction of pressure pulsation for the reciprocating compressor system using finite disturbance theory. J Vib Acoust 2009;131(3):031007.

2. Loh SK, Faris WF, Hamdi M, et al. Vibrational characteristics of piping system in air conditioning outdoor unit. Sci China Technol Sci 2011;54(5):1154–68.

3. Nakamura Tomomichi, Kaneko Shigehiko. Flow induced vibrations: classifications and lessons from practical experiences. Elsevier; 2008.

4. Trebunˇa F, Šimcˇák F, Hunˇady R, et al. Identification of pipes damages on gas compressor stations by modal analysis methods. Eng Failure Anal 2013;27:213–24.

5. Liu B, Feng J, Wang Z, et al. Attenuation of gas pulsation in a reciprocating compressor piping system by using a volume-choke-volume filter. J Vib Acoust 2012;134(5):051002.

6. Dang Xiqi, Chen Shouwu. Gas pulsation and vibration in reciprocating compressor piping system. Xi'an, China: Xi'an Jiaotong University Press; 1984 [in Chinese].

7. Ashrafizadeh Hossein, Karimi Mohsen, Ashrafizadeh Fakhreddin. Failure analysis of a high pressure natural gas pipe under split tee by computer simulations and metallurgical assessment. Eng Failure Anal 2013;32:188–201.

8. Bhatia KG. Foundations for industrial machines-handbook for practising engineers. New Delhi: D-CAD Publishers; 2009.

9. Schrötter M, Trebunˇa F, Hagara M, et al. Methodology for experimental analysis of pipeline system vibration. Procedia Eng 2012;48:613–20.

10. Giampaolo Tony. Compressor handbook: principles and practice. Georgia: The Fairmont Press, Inc.; 2010.

11. Sukaih N. A practical, systematic and structured approach to piping vibration assessment. Int J Press Vessels Pip 2002;79(8):597–609.

12. Liu G, Li S, Li Y, et al. Vibration analysis of pipelines with arbitrary branches by absorbing transfer matrix method. J Sound Vib 2013;332(24):6519–36.

13. Cyklis P. Experimental identification of the transmittance matrix for any element of the pulsating gas manifold. J Sound Vib 2001;244(5):859–70.

14. Kim YW, Lee YS. Damage prevention design of the branch pipe under pressure pulsation transmitted from main steam header. J Mech Sci Technol 2008;22(4):647–52.

15. Riedelmeier S, Becker S, Schlücker E. Measurements of junction coupling during water hammer in piping systems. J Fluids Struct 2014;48:156–68.

16. Sierra-Espinosa FZ, García JC. Vibration failure in admission pipe of a steam turbine due to flow instability. Eng Failure Anal 2013;27:30–40.

17. Lieberman Norman P. Troubleshooting natural gas processing wellhead to transmission. Oklahoma: Pennwell Corp.; 1987.

18. ISO 10816-6. Mechanical vibration evaluation of machine vibration by measurements on non-rotating parts Part 6: reciprocating machines with power ratings above 100 kW; 1995.

19. API 618. Reciprocating compressors for petroleum, chemical, and gas industry service. Washington, D.C.: American Petroleum Institute; 2007.

Modelling, Identification and Control of a Calorimeter Used for Performance Evaluation of Refrigerant Compressors

Rodolfo C.C. Flesch and Julio E. Normey-Rico

Departamento de Automac-ao e Sistemas, Universidade Federal de Santa Catarina, 88040-900, Floriano ~ ´polis, SC, Brazil

ABSTRACT

This paper presents the modelling, identification, and control of a calorimeter used for performance evaluation of refrigerant compressors. A phenomenological model of the process is developed and a simple linear model is obtained considering the dynamics near an operating point. Experimental results are also used to identify a simple linear model from real data. As the process exhibits an integrative behavior plus a dead time, a dead-time compensator specially designed for integrating plants is used to control the output temperature of the calorimeter. Robustness and implementation issues are discussed and experimental results in closed loop are presented to demonstrate the good performance of the controller.

INTRODUCTION

Refrigerant compressors are relatively complex products that require a wide variety of tests to be performed for both product development and quality control. Among those tests, the performance test stands out. Its objectives are to measure refrigerating capacity, active power consumption, isentropic efficiency and the coefficient of performance of the compressor. This kind of test is regulated by international standards, such as ISO-917 (ISO, 1989), EN 13771-1 (CEN, 2003), and ANSI ASHRAE 23 (ASHRAE, 2005), which define topological characteristics of the refrigeration system, measurement uncertainty limits, and control requirements.

Three out of the four parameters measured while testing refrigerant compressors are directly dependent on the mass flow rate of refrigerant through the compressor (ISO, 1989). In accordance with the aforementioned standards, all tests shall comprise two different and independent test methods for measuring the mass flow rate, referred as primary and confirming measurements (ASHRAE, 2005, CEN, 2003 and ISO, 1989). The studied test bench uses dry system refrigerant calorimeter as the primary test method and measurement of refrigerant liquid quantity using a flowmeter as the confirming method.

Both methods require several variables to be controlled in order to allow a precise adjustment of the refrigerant conditions at the compressor inlet and outlet (Navarro, Urchueguía, Gonzálvez, & Corberán, 2005). Some of those variables have dynamics that make the controller quite difficult to be tuned, specially the temperature at the calorimeter outlet. Due to the topology of the circuit inside the calorimeter and the distance between the evaporator outlet and the point where the temperature shall be measured to satisfy standard requirements, a considerable dead time is observed in the temperature response. As traditional feedback controllers do not achieve good performance while controlling this kind of plants, typically the controllers are used in manual mode.

However, the performance of the system can be improved by using a predictor structure. These predictor-based controllers are known as dead-time compensators (DTC) and have been applied to many engineering fields (Huzmezan et al., 2002, Mascolo, 2006 and Normey-Rico and Camacho, 2008). The first DTC algorithm appeared in the late 1950s in a paper by Smith (1957) and from then on it has been one of the most widely used algorithms for dead-time compensation in industry (Normey-Rico & Camacho, 2007).

The case discussed in this paper not only presents a significant dead time but also an integrative behavior. This characteristic of the process demands special attention, since, if the traditional Smith predictor (SP) is used, a constant load disturbance will result in a steady-state error. In order to overcome this problem, many solutions have been proposed in literature since 1981 (Åström et al., 1994, Lu et al., 2005, Normey-Rico and Camacho, 1999 and Watanabe and Ito, 1981).

Comparative results presented in Normey-Rico and Camacho (2007) showed that when simple structure and tuning are desired, a simple DTC can be used to obtain a satisfactory compromise between performance and robustness in many applications. The simple DTC proposed by Matausek and Micić (1996) is a good example of this set of controllers and will be used in this paper.

Despite the development of control strategies for processes with significant dead time, it is difficult to find works reporting the use of these algorithms in industrial applications. The use of a DTC with an integrative laboratory plant is presented in Ingimundarson and Hägglund (2001).

In particular, the application of DTC in refrigeration is still incipient. Conventional compressor-type refrigerators usually employ a thermostat for temperature control, which imposes on–off cycles to the compressor (Aprea, Mastrullo, & Renno, 2006). Bi et al. (2000) and Wang, Hang, Zhang, and Bi (1999)present some applications of proportional-integral derivative (PID) controllers to refrigeration, including multivariable and auto-tuning controllers. In the last years fuzzy and neuro-fuzzy controllers have gained popularity in refrigeration applications, as can be seen in Li et al. (2004), Aprea et al. (2006) and Tian, Feng, and Zhu (2008). Sonntag, Devanathan, Engell, and Stursberg (2007) present the application of a hybrid model predictive controller (MPC) to a supermarket refrigeration system and validate the results through simulation and Leducq, Guilpart, and Trystram (2006) use a non-linear MPC for energy saving in a pilot scale refrigeration plant with variable speed compressor. Garcia-Gabin, Zambrano, and Camacho (2009) applied sliding mode predictive control to an air conditioning solar plant, which has a variable time delay and non-minimum phase behavior.

Concerning control strategies applied to performance tests, there is a considerable lack of information, since there are almost no published references. This kind of test bench generally has more accurate instrumentation and better actuators than typical refrigeration systems, what allows better evaluation of the control strategy itself. On the other hand, the ambient loss of such benches is quite low, what imposes an integral mode to the dynamics of the open-loop system.

The aims of this paper are: (i) to develop a phenomenological model of the calorimeter and a simple model for control purposes, (ii) to design and tune a DTC to control its outlet temperature, and (iii) to perform experiments to validate the model and controller. Moreover, the result of the study allows test benches, which could only be operated in manual mode, to be operated in closed-loop mode.

The paper content is split in five sections. After the introduction, in the second section the process to be controlled is described as well as the process identification procedure is presented. In Section 3 the control structure and tuning are addressed. Controller implementation and experimental results are provided inSection 4 followed by conclusions in Section 5.

PROCESS DESCRIPTION AND IDENTIFICATION

This section is dedicated to presenting more details about the process and describing the model identification procedure and results. The real process is located at the Federal University of Santa Catarina(Brazil) and it has the same structure as the test benches used by the Compressors and Cooling Solutions division of Whirlpool Brazil. The studied bench is used for cooperative studies developed between the university and the company in a partnership project context.

It has a non-linear behavior but, as in many processes in industry, the manipulated and controlled variables are kept close to a desired operating point. Thus, for the model identification procedure, a linear model is computed considering small changes around an operating point.

Compressor Performance Test

Performance test is an experimental activity whose objective is to measure four fundamental characteristics of a compressor related to its performance. This kind of test is mainly used for three purposes: research and development (R&D), determination of catalog parameters, and quality control. The test is performed on special benches, which operate as a refrigeration circuit with many controlled variables. In addition, it is possible to measure a series of variables that are not generally monitored in refrigeration systems. So, it is possible to set the operating condition of the compressor, what makes possible the comparison of different compressors. This comparison is done in terms of the compressor capacity of generating mass flow in such a condition (which relates to the capacity of absorbing heat from the cold sink) and in terms of active power consumption (ASHRAE, 2005 and ISO, 1989).

Refrigerating capacity is measured in watts and is calculated as the product of the mass flow rate of refrigerant through the compressor and the difference between the specific enthalpy of the refrigerant at two points in the refrigeration circuit (ISO, 1989). According to the enthalpy determination points that are chosen, it is possible to obtain different interpretations for the refrigerating capacity (Çengel & Boles, 2007). The most common is the heat removal from the cooled environment. The simplest method for determining the mass flow rate in a refrigeration circuit is to measure it directly by using mass flow meters that are available on the market. Equipments that satisfy the measurement uncertainties required by international standards can easily be found. However, as mentioned before, the measurement of refrigerating capacity is required to be done by two different and independent methods (ISO, 1989).

In order to ensure the independency of the methods, it is typical to use one method based on the direct measurement of the mass flow rate and one method based on the heat balance inside a calorimeter. This type of calorimeter is described in the next section.

Calorimeter and Test Bench Description

A general scheme of the test bench is drawn in Fig. 1, indicating its main components and the measurement points of interest for this paper.

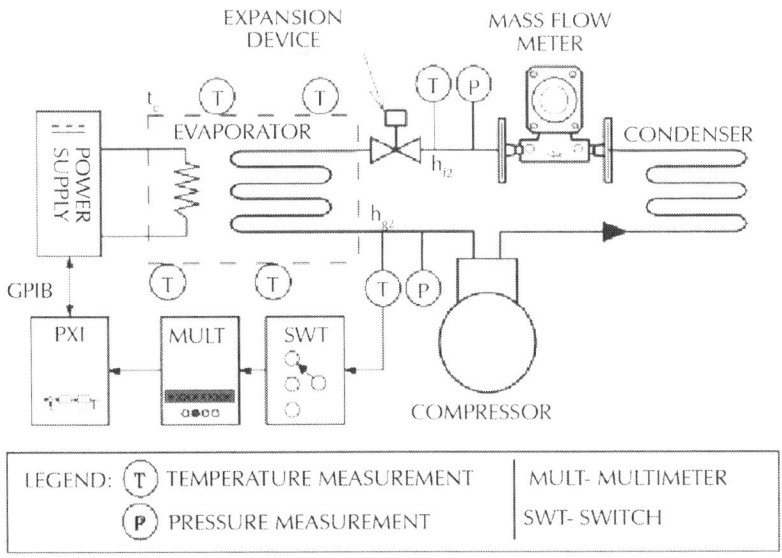

Figure 1: General scheme of the test bench.

A calorimeter is a device used to measure the amount of heat exchanged in a system, so it needs to be a heat-insulated vessel from the environment. The studied test bench uses expanded polystyrene and rock wool as thermal insulators. Concerning performance tests, the calorimeter typically encircles the evaporator and a heater. Considering the case of an electric heater and a dry system refrigerant calorimeter, it is possible to calculate the mass flow rate from Eq. (1) when the system is in steady state:

$$q_{mf} = \frac{\phi_h + F_l(t_a - t_c)}{h_{g2} - h_{f2}}$$

(1)

where $_{qmf}$ is the mass flow rate of refrigerant, $_{\phi h}$ is the electrical power input to the heater, $_{Fl}$ is the heat leakage factor, $_{ta}$ is the average ambient temperature, $_{tc}$ is the average surface temperature of calorimeter, $_{hg2}$

is the specific enthalpy of evaporated refrigerant leaving calorimeter, and $_{hf2}$ is the specific enthalpy of refrigerant liquid entering expansion valve.

Test benches used for performance evaluation are composed by a hydraulic circuit, instruments, controllers and data acquisition systems. The studied test bench consists of a single-stage vapor compression cycle with an electronic expansion device and uses R134a as refrigerant. Temperatures are measured by Pt100 thermoresistors (class A, four wires) connected to a $7\frac{1}{2}$ digit multimeter (PXI-4071 by National Instruments™) through semiconductor switches. Expanded uncertainty (95%) of the measurement chain varies with the measured value, but its upper limit for the operating ranges of the bench temperatures is $\pm 0.5°C$. Pressures are measured by strain-gauge base high accuracy transducers of various types and ranges, but the upper limit for expanded uncertainty of pressure measurements is $\pm 0.2\%$ of each transducer full scale. A direct current (DC) power supply provides the electrical input to the heater and its voltage can be remotely varied from 0 to 200 V over a general purpose interface bus (GPIB). Voltage, current, and power over the heater are measured using a multimeter. The control algorithm and the acquisition software were both developed using LabVIEW and run on a PXI controller (PXI-8186 by National Instruments™).

Calorimeter Modelling

For modelling purposes, a calorimeter can be treated as a perfect heat exchanger. The outlet temperature of a calorimeter at a given time is function of the initial outlet temperature of calorimeter, the average surface temperature of calorimeter, the average ambient temperature (the temperature of the housing where the calorimeter is located), the heat leakage factor, the mass flow rate of refrigerant, the enthalpy of the refrigerant at the inlet and outlet of calorimeter, and the heat input to calorimeter. Due to fluid flow, the effect of power input starts to be felt in the outlet temperature just after a certain amount of time. Thus, a simplified mathematical model for the outlet temperature of a calorimeter can be represented by

$$dt_2(t+L)/dt = K_v[\phi_h(t) + F_l(t_a(t) - t_c(t)) - q_{mf}(t)(h_{g2}(t) - h_{f2}(t))] \tag{2}$$

where $_{t2}$ is the outlet temperature of calorimeter and $_{Kv}$ is the velocity gain. Additionally, t refers to an instant of time and L is the dead time.

From Eq. (2) it is possible to notice that there are other variables than the electrical input to the heater that affect the behavior of the measured variable. Some of these variables were considered constants and some of them were considered as perturbations while modelling the system.

The average ambient temperature typically does not vary during the test and even when it varies its behavior is much slower than the one presented by other variables. Thus, amplitude changes in this variable are quite small and it is reasonable to consider that average ambient temperature is constant. For control purposes, the average surface temperature can be approximated by the average of the temperature of the refrigerant leaving expansion valve ($_{te}$) and the outlet temperature of calorimeter ($_{tc} = (_{te} + _{t2})/2$). Temperature $_{te}$ can be simplified as being the temperature of saturation corresponding to compressor suction pressure, as the refrigerant is biphasic when leaving the expansion device and the expansion device itself is the suction pressure actuator. There is then a direct correlation between $_{te}$ and the suction pressure. Unfortunately, this correlation is not linear and can only be approximated by linear behavior when considering small changes around an operating point.

Mass flow rate is directly related to the compressor operating condition, since it is the product of the volumetric flow rate for the density of the fluid at the same point of reference. As the volumetric displacement per revolution of the compressor is almost constant, the mass flow rate strongly depends on the density of the fluid at the suction line. For the operating conditions of studied refrigerant, density is highly influenced by pressure variations and almost immune to temperature changes. For example, in R134a, the slope for pressure variation around the operating point at the compressor inlet defined by a pressure of 1.1 bar and a temperature of 30°C is about 4kg/(m^3 bar). On the other hand, the slope for temperature changes is about -0.02kg/(m^3 K) at the same operating point. All data shown were obtained from REFPROP 8.0 (Lemmon, Huber, & McLinden, 2007).

The enthalpy of refrigerant entering the evaporator is approximately the same as the enthalpy of refrigerant entering the expansion device (Çengel & Boles, 2007), so their value is defined by the conditions upstream of the expansion device. These conditions vary with refrigerant charge level and the liquid refrigerant reservoir temperature, but none of them are supposed to vary during a test. Thus, the enthalpy of refrigerant entering the evaporator can be considered constant. The enthalpy of refrigerant leaving the evaporator can be determined as a function of pressure and temperature at that point. The evaporator outlet and suction pressures are almost the same, since there is typically a small pressure drop of a few tens of millibars between the two points. The evaporator temperature is the variable of interest for the control system and its variation causes enthalpy variation. Even though it is clear the relation between temperature variation and enthalpy variation, this influence will not be considered for modelling purpose. According to Lemmon et al. (2007), this enthalpy variation is small when compared to the enthalpy variation due to phase change.

As should be clear from previous paragraphs, there is a correlation between calorimeter outlet temperature and suction pressure. To maintain the simplicity of the model, it is assumed that suction pressure variations are much faster than calorimeter outlet temperature dynamics. It is possible, then, to consider a step variation of suction pressure as a step load disturbance, since both the multiplication of mass flow rate by the variation of enthalpy and the multiplication of evaporation temperature by the leakage factor F_l have power units.

Inserting previous modifications in Eq. (2) and grouping the constants, the model can be written as presented in Eq. (3). It shall be noted that a suction pressure variation will change the constant part, but the new value of the constant can be estimated numerically by using measured data:

$$\frac{dt_2(t+L)}{dt} + \frac{K_v F_l}{2} t_2(t) = K_v \phi_h(t)$$
$$-\left\{ K_v \left[\frac{F_l}{2} t_e + q_{mf}(h_{g2} - h_{f2}) - F_l t_a \right] \right\}$$

(3)

The term in square brackets has power unit and can be interpreted as the heat removal caused by evaporation and exchanges with the environment. Representing this heat removal as ϕ_r, shifting the model L seconds and integrating Eq. (3), it results in

$$t_2(t) + \frac{K_v F_l}{2} \int_0^t t_2(\tau - L)\, d\tau = K_v \int_0^t [\phi_h(\tau - L) - \phi_r]\, d\tau$$

(4)

Applying Laplace transform in Eq. (4) and considering null initial conditions and $\Delta\phi(t)$ as being $\phi_h(t)-\phi_r$, it results in Eq. (5), where T(s) and $\Delta\phi(s)$ are the Laplace transforms of $_{t2}(t)$ and $\Delta\phi(t)$, respectively:

$$T(s) + \frac{K_v F_l}{2} \frac{T(s)e^{-Ls}}{s} = K_v \frac{\Delta\phi(s)e^{-Ls}}{s}$$

(5)

From Eq. (5) it is easy to derive Eq. (6),that is the transfer function of the model ($_{Pm}(s)$), by multiplying both sides of the equation by two times the Laplace variable s   and grouping the terms in T(s):

$$P_m(s) = \frac{T(s)}{\Delta\phi(s)} = \frac{2K_v e^{-Ls}}{2s + K_v F_l e^{-Ls}}$$

(6)

It must be noted that there are other dynamics which have not been considered in this model due to simplifications. It is common to model these dynamics using a low-order system in series with the proposed model (Normey-Rico & Camacho, 2007), as shown in Eq. (7), where $_{Tu}$ is the equivalent time constant of the unmodelled dynamics:

$$P_s(s) = P_m(s) \frac{1}{T_u s + 1}$$

(7)

In this paper, those dynamics have not been included in the model used for controller design, but they have been taken into account in the robustness analysis of Section 3. From experimental results, $_{Tu}$was estimated to be 40 s.

Process Identification And Validation

Based on analytical modelling, parameters were identified to determine the model for controller design purposes. The first parameter to be identified was heat leakage factor. Identification was done in accordance with the heat loss method presented in ISO (1989). The ambient temperature was maintained constant to within ± 1 k and heat was supplied to maintain refrigerant temperature 15 k above the ambient temperature. Electric heat was also maintained to within ±1%. After thermal equilibrium was established, successive readings of temperature and electrical power input to the heater were recorded. Heat leakage factor was determined using experimental values and Eq. (8).

$$F_l = \frac{\phi_h}{t_c - t_a}$$

(8)

Estimated heat leakage factor for the studied calorimeter is 0.52w/k. However, during the experiments, the ambient temperature of calorimeter is controlled and maintained to within $\pm 3K$ of the average surface temperature of calorimeter. Considering the worst case, deviation of $3K$, the power loss to environment is 1.56 w, which is <1% of the normal power used for heating. Furthermore, $_{Kv}$ is typically small in magnitude (<0.01K/(s W)), what results in a system with a pole near the origin. If compared to the closed-loop desired dynamics, the open-loop model can be approximated by an integrator from the point of view of the controller. The use of this information allows suppressing the term $_{KvFl}e^{}Ls$ in Eq. (6) and the system becomes a pure integrator with dead time L   and velocity gain $_{Kv}$:

$$P_m(s) = \frac{K_v e^{-Ls}}{s}$$

(9)

In the experimental identification tests, $\Delta\phi$ was excited with different pulses and $_{Kv}$ and L were determined experimentally from the open-loop response, as is shown in Fig. 2.

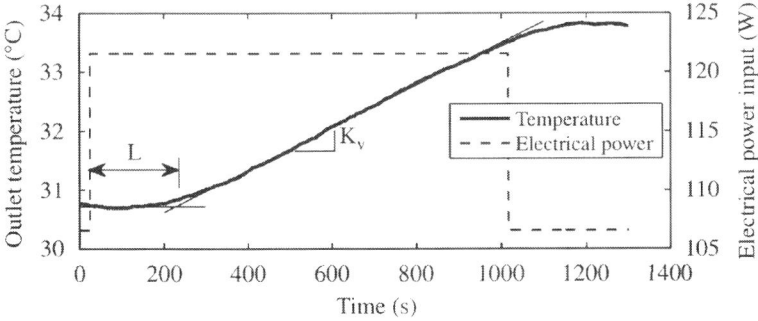

Figure 2: Model identification.

Using several experiments near the operating point, an average nominal model ($_{Pn}(s)$) was obtained with L_n =200s and K_v=2.4×10⁻⁴K/(sW), as shown in Eq. (10). In addition, it was observed that the maximum dead-time estimation error Δ_{Lmax} is 40 s:

$$P_n(s) = \frac{T(s)}{\Delta\phi(s)} = \frac{2.4 \times 10^{-4}e^{-200\,s}}{s}$$

(10)

THE PROPOSED CONTROLLER

The simple structure of the DTC presented in Mataušek and Micić (1996) is used here to derive the process temperature controller. The proposed controller is tuned using a robustness analysis and it also considers the effect of the saturation in the control action.

DTC Structure and Nominal Tuning

The control scheme can be seen in Fig. 3, where $_{Pn}(s)$ is the process model and $_{Gn}(s)$ is the dead-time free model. The idea here is that when step disturbances are considered, the difference between model and process outputs in steady state is an estimation of the amplitude of the load disturbance q(t). Thus, using M(s), this estimated value is introduced in the loop to eliminate the effect of the disturbance. As M(s) acts as a feedback controller, it must be tuned to achieve closed-loop stability and also an adequate transient of the disturbance rejection response. Moreover, the primary controller C(s) is used to define the set-point response of the closed-loop system.

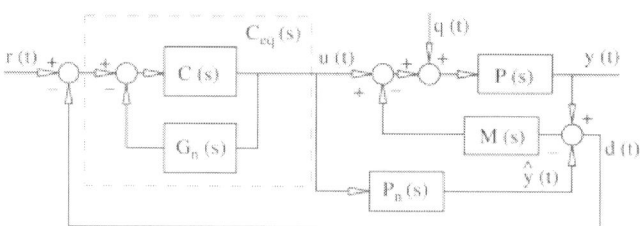

Figure 3: Modified Smith predictor proposed by Mataušek and Micić (1996).

The most simple solution for this problem is to use model $P_n(s) = K_v e^{-L_n s}/s$ and a simple gain in each controller (Mataušek & Mici , 1996), i.e. $C(s) = K_c$ and $M(s) = K_0$. In this case, considering perfect modelling and null load disturbance ($q(t) = 0$), the prediction of the output, given by $\hat{Y}(t)$, is equal to the output $y(t)$, so the prediction error $d(t)$ is null. In this case, the Laplace transform of the control signal is $U(s) = C_{eq}(s)R(s)$ and the Laplace transform of the output is $Y(s) = P(s)U(s)$, where $R(s)$ is the Laplace transform of $r(t)$. Given that $C_{eq}(s) = C(s)/(1+C(s)G_n(s)) = K_c s/(s + K_c K_v)$ and $P(s) = K_v e^{-L_n s}/s$, the closed-loop transfer function is the one presented in Eq. (11), where T_r is the desired closed-loop time constant:

$$\frac{Y(s)}{R(s)} = \frac{e^{-L_n s}}{T_r s + 1}, \quad T_r = \frac{1}{K_c K_v}$$

(11)

That is, T_r can be used to define the speed of the set-point response and the controller gain is $K_c = 1/(T_r K_v)$. To complete the tuning of the controller, $K_0 = 1/(2 K_v L_n)$ is computed imposing a phase margin of approximately $\phi_m = 60°$ to the inner loop (Mataušek & Mici , 1996). With these parameters, in the nominal case, the transfer function relating the output $Y(s)$ to the disturbance $Q(s)$ is Eq. (12), which is, as expected, a stable transfer function with zero static gain (Normey-Rico & Camacho, 2008):

$$\frac{Y(s)}{Q(s)} = \left[1 - \frac{e^{-L_n s}}{T_r s + 1}\right]\left[\frac{K_v e^{-L_n s}}{s + K_0 K_v e^{-L_n s}}\right]$$

(12)

Robustness Analysis

For the analysis of robustness it is considered that the plant model $P_n(s)$ is different from the process $P(s)$, $P(s) = P_n(s)[1 + \delta P(s)]$, thus the condition for robust stability (Morari & Zafiriou, 1989) is presented in Eq. (13), where δP is a multiplicative description of the modelling errors, dP is an upper bound of δP, and ω is the angular frequency. The derivation of the upper bound of the modelling errors is based on a worst case analysis that uses the Nyquist theorem. Further details can be found in (Normey-Rico & Camacho, 2008):

$$|\delta P(j\omega)| < dP(\omega) = \frac{|j\omega(j\omega + K_c K_v)(2L_n j\omega + e^{-j\omega L_n})|}{|1 - (2L_n K_v K_c + 1)\omega^2 + j\omega K_v K_c|}$$

(13)

As shown in Normey-Rico and Camacho (2002) it is possible to tune parameter T_r using an estimation of the maximum value of the model uncertainties defined by T_u (the equivalent time constant of the unmodelled dynamics) and $\Delta L = L - L_n$:

$$\frac{2L_n T_r}{2L_n + T_r} > \Delta L + T_u$$

(14)

A parameter β can be defined for robust stability ($\beta = \Delta_{Lmax} + T_u$) or for robust performance ($\beta = 2(\Delta_{Lmax} + T_u)$) and the tuning of T_r shall be done in accordance with Eq. (15), as shown in (Normey-Rico & Camacho, 2007):

$$T_r > \frac{\beta}{1 - \frac{\beta}{2L_n}}$$

(15)

Fig. 4 shows δP for the case analyzed in this paper and the maximum estimated values of T_u and ΔL, as well as dP for two values of T_r: one chosen for robust stability ($T_r = 100s$) and the other for robust performance ($T_r = 300s$). As can be seen, the proposed tuning gives the expected results. To show the effect of T_u and Δ_{Lmax} on the modelling error, δP is also shown in dotted line for the case $T_u = 10s$ and $\Delta L_{max} = 10s$.

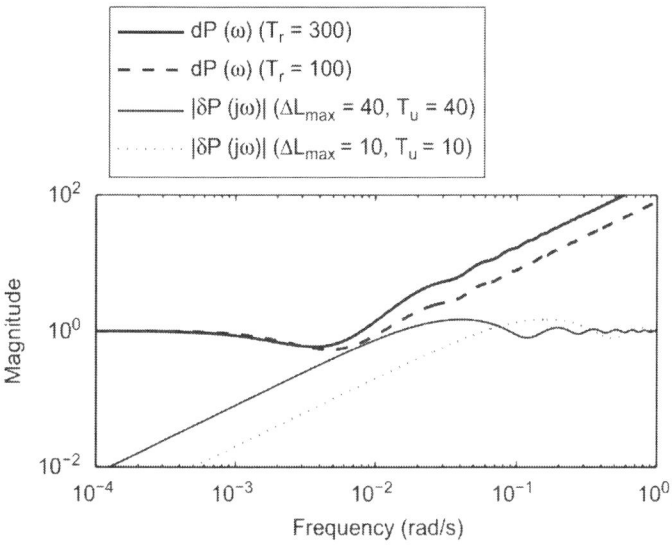

Figure 4: Robustness analysis for the proposed controller.

Including Saturation and Set-Point Weighting

The main advantage of this controller is that it has only three tuning parameters: $_{Kv'\ Tr}$ and $_{Ln'}$ like a PID controller. One of its disadvantages, if compared to more elaborated DTC, is that $_{Tr}$ affects the disturbance rejection and the set-point responses and, therefore, it does not allow set-point and disturbance responses to be decoupled. To overcome this problem without increasing the controller complexity, a set-point filter is used in the DTC (see Fig. 5) to avoid the overshoots in some particular situations where big changes in the desired temperature are considered.

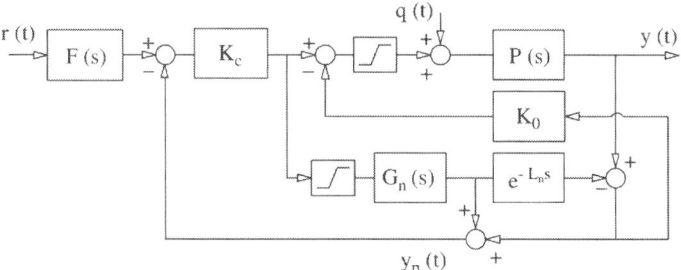

Figure 5: DTC including saturations and a reference filter.

Finally, to take into account the actuator saturation on the controller, a modified DTC structure is used in real implementation, as the one shown in Fig. 5. Saturation is included in the model of the plant at the input of the fast model, maintaining all the other blocks.

The inclusion of the saturation at this point keeps internal stability and solves the integrator windup problem for reference tracking. Simulation results showed that this anti-windup scheme produces good results when the control signal saturates, as can be seen in Fig. 6. The main idea is to use the same control signal in the real plant and in the model by including the saturation information explicitly in the model, avoiding, thus, unnecessary oscillations caused by disparity between the real control signal and the control signal of the model.

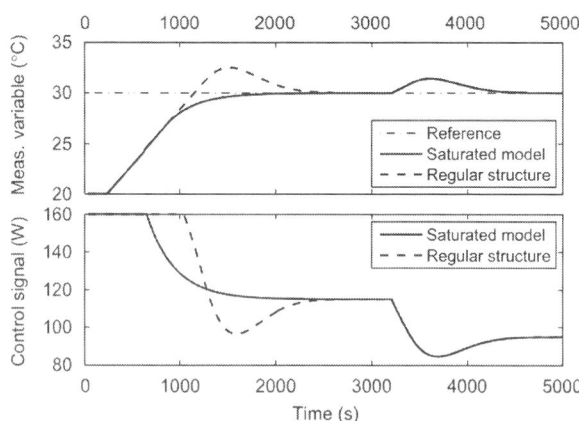

Figure 6: Simulation results for system response with anti-windup scheme.

Note that in the nominal case (no modelling error and no disturbances), two fundamental properties of the DTC hold for this structure: (i) the dead time is eliminated from the main feedback loop and (ii) the signal prediction $_{yp}(t)$ coincides with the system output at t+L.

CONTROLLER IMPLEMENTATION AND EXPERIMENTAL RESULTS

For the process model presented in Section 2, L_n=200s , K_v=2.4 × 10⁻⁴K/(sW), and the unmodelled dynamics and dead-time error are estimated using average values from the experiments giving T_u=40s and

$\Delta L_{max} = 40s$. Computing β=2(Δ_{Lmax}+$_{Tu}$)=160, $_{Tr}$ must be higher than 267 s for a robust behavior. As a settling time of the closed-loop response of about 15 min (900 s) is adequate for the process, T_u=300s has been chosen. Using the proposed tuning rule for the internal loop, $_{K0}$=10.4.

A discrete version of the proposed dead-time compensator is used in the real experiments, thus the obtained controller was discretized using a sampling time T_s=300s. In this case, only the process model$_{Pn}$(s) has to be discretized and, as shown in Normey-Rico and Camacho (2007), this must be done using a zero order hold:

$$P_n(z) = \frac{K_v T_s}{z - 1} z^{-L_n/T_s}$$

(16)

A first-order discrete set-point reference filter with unitary static gain was designed as $_{Fr}$(z)=0.05/(z-0.95). Experimental results showed that it was necessary for step references that deviates more than 2°C from the operating point. The final block diagram representation for implementation is presented in Fig. 7.

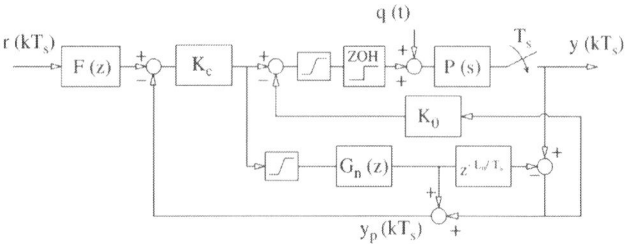

Figure 7: Discrete version of DTC used for implementation.

Experimental results for a step reference of 32.2°C applied at t=0s are presented in Fig. 8. On all test results for this condition, temperature was maintained to within ±0.1°C of the desired value in steady state. A load disturbance was applied at t=1400s to the system by changing the suction pressure of the compressor under test. It is clearly demonstrated that the proposed DTC is able to produce good results in both the set-point and load disturbance responses. Note that system was not in equilibrium on the beginning of operation due to test conditions, but it can be seen that the dead time observed for reference tracking has the same magnitude as the one observed for disturbance rejection. In addition, it shall be noted that the signal-to-noise ratio in temperature measurement is very high, so noise does not impose additional difficulties on the control design.

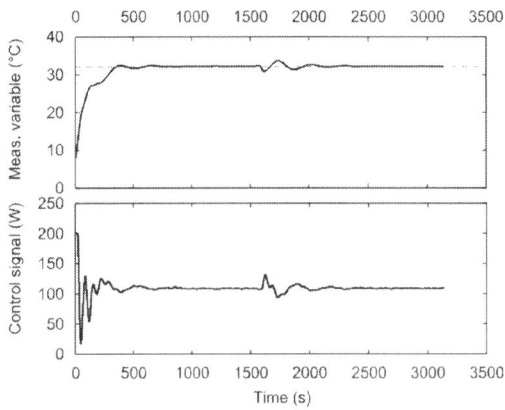

Figure 8: Closed-loop responses for a typical experiment using a compressor whose nominal refrigerating capacity is 115 W.

The robust behavior of the controller was experimentally analyzed by testing different models of compressors under different conditions. Fig. 9 presents the time response of the controlled variable observed when testing a larger compressor, which imposes a greater mass flow rate. This compressor was tested with a set point of 23.5°C (note that y-axis values are different from Fig. 8). For testing this compressor, the only modification in control implementation was the change of linearization point of power input to a value near the heat removal by means of evaporation. Again, a load disturbance was applied to the system by changing the suction pressure of the compressor under test and the control system was able to drive the controlled variable back to set point.

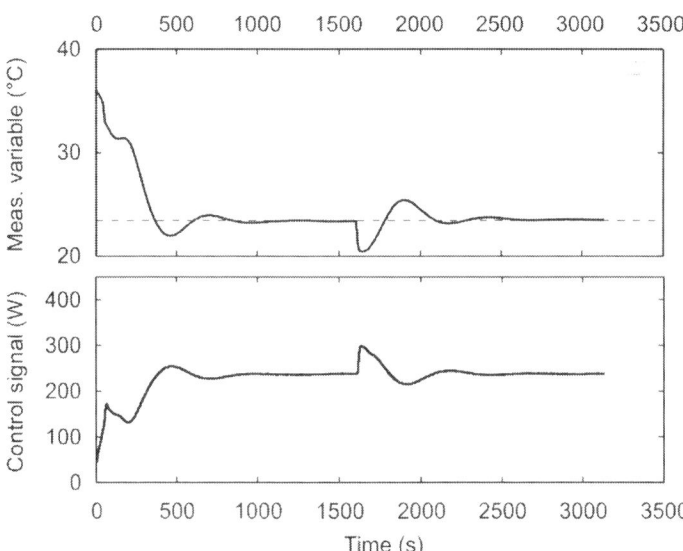

Figure 9: Closed-loop response for a different model of compressor whose nominal refrigerating capacity is 240 W.

Fig. 10 shows the response of the closed-loop system when testing the same compressor used for Fig. 8experiments, but subject to another condition that differs from that used for modelling purposes. It is noticeable that the performance is degraded in comparison with the nominal case, but the closed-loop system remains stable and follows the set point.

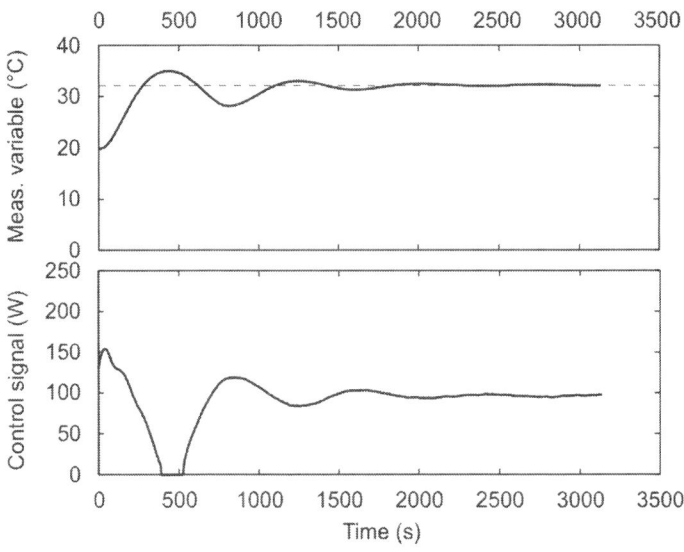

Figure 10: Closed-loop response for different conditions.

CONCLUSIONS

The paper has focused on presenting the modelling and identification of a calorimeter used for performance evaluation of refrigerant compressors. Moreover, the application of a DTC to control the outlet temperature of this process was analyzed. Real experiments using DTC to control industrial integrative plus dead-time processes are rarely seen in literature.

Experimental results showed that the studied controller performs well both for set point changes and disturbance rejection. In addition, it has a simple tuning procedure with only three parameters, being two of them characteristics that can be determined from the open-loop response.

Besides being an interesting plant from the point of view of control, this kind of test bench is much used by compressor manufacturers. The result of the study allows test benches, which are typically operated in manual mode, to be operated in closed-loop mode. Automation leads to improvements in test repeatability and time consumption. These gains reflect more trustable information for R&D, catalog data, and product approval. In addition, those benches are very expensive and automatic control increases the productivity, allowing more compressors to be tested at the same bench by unit of time. Furthermore, the paper content can be applied not just to performance evaluation benches but also to refrigeration systems, such as refrigerators and freezers.

ACKNOWLEDGMENTS

The authors would like to express sincere gratitude to the Laboratory for Instrumentation and Test Automation Applied to Refrigeration (LIAE) of the Mechanical Department of the Federal University of Santa Catarina, particularly to Professor Carlos Alberto Flesch for the technical and instrumental support. The authors also thank the Compressors and Cooling Solutions division of Whirlpool Brazil, especially M.Eng. Maikon Ronsani Borges, for financial and technical support. Finally, financial support from the Brazilian funding agencies CNPq and FINEP is gratefully acknowledged.

REFERENCES

1. Aprea, C., Mastrullo, R., & Renno, C. (2006). Experimental analysis of the scroll compressor performances varying its speed. Applied Thermal Engineering, 26(10), 983–992.

2. ASHRAE (2005). ANSI/ASHRAE 23: Methods of testing for rating positive displacement refrigerant compressors and condensing units.

3. Astr ° om, K. J., Hang, C. C., & Lim, B. C. (1994). A new Smith predictor for controlling ¨ a process with an integrator and long dead-time. IEEE Transactions on Automatic Control, 39(2), 343–345.

4. Bi, Q., Cai, W.-J., Wang, Q.-G., Hang, C.-C., Lee, E.-L. Sun, Y., et al. (2000). Advanced controller auto-tuning and its application in HVAC systems. Control Engineering Practice, 8(6), 633–644.

5. CEN (2003). EN 13771-1: Compressors and condensing units for refrigeration – Performance testing and test methods. Part 1: Refrigerant compressors.

6.   C- engel, Y. A., & Boles, M. A. (2007). Thermodynamics: An engineering approach (6th ed.). New York: McGraw-Hill Higher Education.

7.   Garcia-Gabin, W., Zambrano, D., & Camacho, E. F. (2009). Sliding mode predictive control of a solar air conditioning plant. Control Engineering Practice, 17(6), 652–663.

8.   Huzmezan, M., Gough, W. A., Dumont, G. A., & Kovac, S. (2002). Time delay integrating systems: a challenge for process control industries. A practical solution. Control Engineering Practice, 10(10), 1153–1161.

9.   Ingimundarson, A., & Hagglund, T. (2001). Robust tuning procedures of dead-time ¨ compensating controllers. Control Engineering Practice, 9(11), 1195–1208.

10.   ISO (1989). ISO 917: Testing of refrigerant compressors.

11.   Leducq, D., Guilpart, J., & Trystram, G. (2006). Non-linear predictive control of a vapour compression cycle. International Journal of Refrigeration, 29(5), 761–772.

12.   Lemmon, E. W., Huber, M. L., & McLinden, M. O. (2007). NIST standard reference database 23: Reference fluid thermodynamic and transport properties (REFPROP) version 8.0. Standard reference data program.

13.   Li, X., Chen, J., Chen, Z., Liu, W., Hu, W., & Liu, X. (2004). A new method for controlling refrigerant flow in automobile air conditioning. Applied Thermal Engineering, 24(7), 1073–1085.

14.   Lu, X., Yang, Y.-S., Wang, Q.-G., & Zheng, W.-X. (2005). A double two-degree-offreedom control scheme for improved control of unstable delay processes. Journal of Process Control, 15(5), 605–614.

15.   Mascolo, S. (2006). Modeling the internet congestion control using a Smith controller with input shaping. Control Engineering Practice, 14(4), 425–435.

16.   Matausek, M. R., & Micic ˇ ´, A. D. (1996). A modified Smith predictor for controlling a process with an integrator and long dead-time. IEEE Transactions on Automatic Control, 41(8), 1199–1203.

17.   Morari, M., & Zafiriou, E. (1989). Robust process control. Englewood Cliffs, NJ: Prentice-Hall.

18. Navarro, E., Urchueguı́a, J. F., Gonzáĺvez, J., & Corberán, J. M. (2005). Test results of performance and oil circulation rate of commercial reciprocating compressors of different capacities working with propane (R290) as refrigerant. International Journal of Refrigeration, 28(6), 881–888.

19. Normey-Rico, J. E., & Camacho, E. F. (1999). Robust tuning of dead-time compensators for processes with an integrator and long dead-time. IEEE Transactions on Automatic Control, 44(8), 1597–1603.

20. Normey-Rico, J. E., & Camacho, E. F. (2002). A unified approach to design dead-time compensators for stable and integrative processes with dead-time. IEEE Transactions on Automatic Control, 47(2), 299–305.

21. Normey-Rico, J. E., & Camacho, E. F. (2007). Control of dead-time processes. London: Springer.

22. Normey-Rico, J. E., & Camacho, E. F. (2008). Dead-time compensators: A survey. Control Engineering Practice, 16(4), 407–428. Smith, O. (1957). Closed control of loops with dead time. Chemical Engineering Progress, 53, 217–219.

23. Sonntag, C., Devanathan, A., Engell, S., & Stursberg, O. (2007). Hybrid nonlinear model-predictive control of a supermarket refrigeration system. In 16th IEEE international conference on control applications (pp. 1432–1437). Singapore: IEEE.

24. Tian, J., Feng, Q., & Zhu, R. (2008). Analysis and experimental study of MIMO control in refrigeration system. Energy Conversion and Management, 49(5), 933–939.

25. Wang, Q.-G., Hang, C.-C., Zhang, Y., & Bi, Q. (1999). Multivariable controller autotuning with its application in HVAC systems. In Proceedings of the 1999 American control conference, Vol. 6 (pp. 4353–4357). San Diego, CA: IEEE.

26. Watanabe, K., & Ito, M. (1981). A process-model control for linear systems with delay. IEEE Transactions on Automatic Control, 26(6), 1261–1269.

5

Calculation of Temperature Distributions in the Rotors of Oil-Injected Screw Compressors

S.H. Hsieh[a], Y.C. Shih[a], Wen-Hsin Hsieh[a], F.Y. Lin[b], and M.J. Tsai[b]

[a]Department of Mechanical Engineering, National Chung Cheng University, 168 University Road, Ming-Hsiung, Chia-Yi 621, Taiwan, ROC

[b]Fu Sheng Industrial Co., Ltd., No. 60, Sec. 2, Guangfu Rd., Sanchong City, Taipei County, Taiwan, ROC

ABSTRACT

A mathematical model and a calculation procedure are proposed in this study to efficiently calculate the temperature distributions in the male and female rotors of the oil-injected screw compressors. The solution of the transient heat conduction problem of the rotors, which is subject to a periodic convective boundary condition and five steady boundary conditions, is obtained by solving the set of Helmholtz equations derived from the partial differential equations for transient heat conduction without internal heat. During the solving process, the periodic convective and the five steady boundary conditions are calculated using the six empirical constants together with the ideal convective heat-transfer coefficient and the ideal steady heat fluxes. The six empirical constants are determined by minimizing the difference between the calculated and measured temperatures of the rotors. The average errors of calculated temperatures at three locations of each of the male and female rotors are 5.45% and 4.85%, respectively. Results indicate that the heat is mainly transferred from the bearings installed on the outlet shaft of the rotors into the screws. Then, the heat is transferred from the screws to the compressed air and the bearings installed on the inlet shaft of the rotors. The temperature gradient in the axial direction is different at different positions along the screw. The results of the sensitivity analysis show that the three empirical constants that are used to calculate the heat convection between the screw and the compressed air, the heat transfer between the rotor and the bearing at the inlet shaft of the rotor and the heat transfer between the rotor and the bearing at the outlet shaft of the rotor, have a stronger impact on the outputs of the mathematical model than the other three empirical constants. The reduced mathematical model, using only the three empirical constants with the strongest impact, can be used in the studies of the temperature distribution in the rotor. The calculated temperature distributions in the screws can be used to estimate the approximate thermal deformation of the screws and thereby improve the screw profiles and the design of the oil-injected screw compressor.

INTRODUCTION

An oil-injected screw compressor compresses air within the compression chamber formed by the casing and the screws of the male and female rotors of the compressor. As the rotors rotate, the volume of the compression chamber changes and thereby compresses the air. To cool the compressor and the compressed air and to lubricate the rotating meshing rotors, lubricant oil is injected into the compression chamber at the proper position during a compression cycle. Currently, the volumetric efficiency of screw compressors is higher than 90% due to the well-controlled clearances of the internal leakage paths achieved with modern precision-machining techniques [1]. When designing an oil-injected screw compressor, one of the major design goals is to maximize the volumetric and the isentropic efficiencies [2],[3], [4] and [5]. Thermodynamic analysis is usually performed at first, followed by the analysis of the loads exerted on the rotors. With the results of the thermodynamic and load analysis, the design of the screw profiles of the male and female rotors and of the geometric structure of the case are performed and optimized. Stosic et al. [5] compiled a list of all 5 × 6 screw profiles that had been generated since 1967 and the procedures to calculate these profiles.

Studies on the compression processes of oil-injected screw compressors can be classified into two categories, lumped parameter analysis and finite element/difference analysis. In the first category, a lumped parameter system is adopted to analyze the performance of the oil-injected screw compressor. The properties of the compressed air inside the compression chamber are considered to be uniform, but time dependent, and their changes are described by the conservation of mass and energy. The calculated volumetric efficiency, isentropic efficiency and pressure curve are verified by experimental results.

Using lumped parameter analysis, Stosic et al. [6] established the formulations to describe the flying path of oil that is injected into the compression chamber. A parametric study was performed to investigate the oil-injection parameters on the performance of the compressor. Fujiwara and Osada [7] obtained the correlations of the heat-transfer coefficients between the compressed air and oil inside an oil-injected screw compressor. Fleming et al. [8], [9] and [10] studied the leakage paths inside the screw compressor and noted that the clearances of the leakage paths affect the performance of the screw compressor. Wu et al. [11], in a two-phase flow analysis, considered the relative motion between the compressed air and oil and calculated the mass flow rates at the leakage paths and at the outlet. The mathematical model was verified by experimental results. Seshaiah et al. [12] calculated the mass flow rates of the compressed air and the oil through the clearances of the leakage paths. They studied not only the effects of the clearance of the leakage paths on the performance of the compressor, but also the effects of the mass ratio of the compressed air to oil on the temperature–time histories of the compressed air and oil. Stosic et al. [5] and Lee et al. [13] used the calculated properties of the compressed air that were obtained through a lumped parameter analysis to calculate the average axial radial forces exerted on the bearing. The effects of the input power, the loads exerted by the compressed air and the contact forces between the screws on both the male and female rotors were considered in the analysis.

The major advantage of using a lumped parameter analysis is that the accurate calculated performance of the oil-injected screw compressor can be obtained in a particularly short time, usually in a few seconds. The designers of the compressor can obtain the design information quickly and efficiently, but the detailed phenomena, such as the temperature distribution and the thermal deformation in the rotors, are not available to the designers.

In the second category of study, the thermal and/or solid properties of the structure of the screw compressor and the compressed air are calculated by the finite element method and the finite difference method. Kovacevic et al. [14], [15] and [16] developed a procedure to generate variable meshes that with time within the compression chamber. The variable meshes were shown to improve the accuracy of the calculations for the flow field of the compressed air. In addition to the flow field, the temperature distribution in the rotors, the deformations of rotors and the stresses in the solid parts of the screw compressor were also calculated. One of the advantages of the second category of study is that the results of the flow field, the temperature distribution, the deformations and the stresses are obtained at the same time by solving the applied mathematical model [14], [15] and [16]. Detailed information is available for the design optimization of the screw compressor. However, the calculation time required to obtain the steady-state convergent results, which describe the transient phenomena of the oil-injected screw compressor during an entire compression cycle, is usually huge.

In view of the above-mentioned advantages and disadvantages of the two categories of study, a new procedure and a mathematical model are developed in this study to calculate efficiently the compression processes of the air and the temperature distributions in the rotors. With this new methodology, it is possible to conduct a coupled analysis of the screw profiles and thermo-fluidic properties within a reasonable time frame. The solution of the transient heat conduction problem of the rotors, which is subject to a periodic boundary condition and five steady boundary conditions, is obtained by solving the set of Helmholtz equations that are derived from the partial differential equations for transient heat conduction without internal heat [17]. The transient three-dimensional problem is deduced to a steady three-dimensional problem and the calculation time is therefore significantly reduced.

During the solving process, the periodic convective and the five steady boundary conditions are calculated. To determine the periodic convective boundary condition, the convective heat-transfer coefficient is calculated as the product of an empirical constant and an ideal convective heat-transfer coefficient. The ideal convective heat-transfer coefficient is determined by the correlation proposed by Gnielinski [18] using the properties of the compressed air that are obtained with the lumped parameter analysis developed by the authors [19]. To determine the five steady boundary conditions, each of the heat fluxes at five boundaries is calculated as the product of an empirical constant and an ideal heat flux. The ideal heat fluxes are calculated as the bearing power loss [20] and [21] and the frictional power losses of oil at the clearances of the leakage paths [22]. In addition, the six empirical constants are determined by minimizing the difference between the calculated and measured temperatures of the rotors.

The mathematical model is verified by the experimental results. The temperature distributions in the male and female rotors, the heat conduction in the rotors and the thermal deformation are discussed based on the calculated results. A sensitivity analysis is also implemented to determine the effects of the six empirical constants on the temperature distribution in the rotor. The results of sensitivity analysis show that reducing the number of the empirical constants is feasible.

MATHEMATICAL MODEL

In this study, the mathematical model describes the temperature distributions in the steadily operating male and female rotors of a commercial oil-injected screw compressor. The geometric parameters of the compressor are listed in Table 1. The 5 × 6 screw profiles are generated by Normal–Rack Generation Method developed by Wu and Fong [23]. The solid models, established based on the screw profiles, of the male and female rotors can be separated into three parts: the screw, the inlet shaft of the rotor and the outlet shaft of the rotor, as shown in Fig. 1. The additional geometric parameters, including the volume of the compression chamber, the lengths of the leakage paths and the areas of the inlet, outlet and the oil-injected ports are also calculated based on the screw profiles.

Table 1: The geometrical parameters of the male and female screws

Parameters	Male screw	Female screw
Number of teeth	5	6
Outer diameter (m)	0.113	0.090
Length of screw (m)	0.183	0.183
Lead (m/(2ϖ))	0.221	0.265
Clearance of the inlet end of the screw (m)	3.7×10^{-4}	3.7×10^{-4}
Clearance of the outlet end of the screw (m)	4×10^{-5}	4×10^{-5}
Clearance of the sealing line of the screw (m)	6×10^{-5}	6×10^{-5}

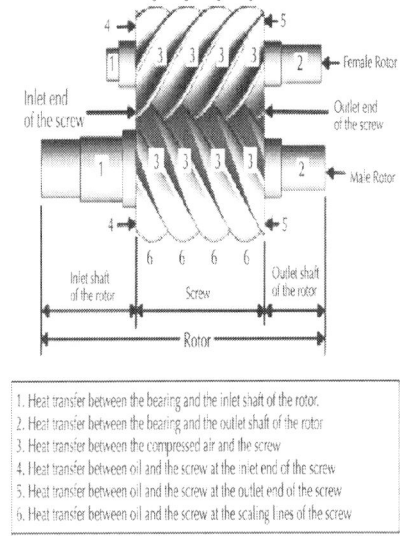

1. Heat transfer between the bearing and the inlet shaft of the rotor.
2. Heat transfer between the bearing and the outlet shaft of the rotor
3. Heat transfer between the compressed air and the screw
4. Heat transfer between oil and the screw at the inlet end of the screw
5. Heat transfer between oil and the screw at the outlet end of the screw
6. Heat transfer between oil and the screw at the sealing lines of the screw

Figure 1: The solid models of the male and female rotors and the six boundary conditions of the mathematical model.

When the screw compressor operates steadily, the temperature distributions in the male and female rotors are affected by six heat-transfer boundary conditions, which are shown schematically in Fig. 1. The first boundary condition is the heat-transfer between the rotor and the bearing at the inlet shaft of the rotor; the second is the heat-transfer between the rotor and the bearing at the outlet shaft of the rotor; the third is the convective heat-transfer between the screw and the compressed air; the forth is the heat-transfer between the screw and oil at the inlet end of the screw; the fifth is the heat-transfer between the screw and oil at the outlet end of the screw; the sixth is the heat-transfer between the screw and oil at the sealing line. Among the six considered boundary conditions, the convective heat-transfer between the screw and the compressed air is transient and periodic, and the other five heat-transfer boundary conditions are assumed to be steady [21] and [22]. In addition, the male and female rotors are designed to contact with each other along a line (the contact line). The contact area between the rotors is, therefore, negligibly small when compared with the contact area between the compressed air and the rotors. Furthermore, the temperature difference between the two rotors is also smaller than that between the compressed air and rotors. Therefore, the heat-transfer between the rotors is much smaller than the convective heat-transfer between the compressed air and rotors, and is assumed to be negligible.

Heat Conduction Problem Involving Periodic Boundary Condition

The aim of this study is to investigate the temperature distributions in the male and female rotors, which operate steadily and are subjected to one periodic and five steady boundary conditions. The transient, three-dimensional heat conduction equation is shown in Eq. (1) for a solid without internal heat generation. The parameter α is the thermal diffusivity of the rotor. In this study, the male and female rotors are both made of cast iron that has a thermal diffusivity of 1.67×10^{-5} m²/s [24].

$$\nabla^2 T = \frac{1}{\alpha} \frac{\partial T}{\partial t}$$

(1)

Among the six heat-transfer boundary conditions, the heat convection between the screw and the compressed air is the periodic one and is described by Eq. (2).

$$k\frac{\partial T}{\partial n} = h \cdot (T_g - T)$$

(2)

In Eq. (2), k is the heat conductivity of the rotor [51 W/(m K)] [24], n is the unit vector normal to the boundary, T is the temperature of the screw at the boundary, and T_g is the temperature of the compressed air, which is periodic. The parameter h is the heat-transfer coefficient. Since the thickness of the oil film, which covers screws, is very small, the influence of the oil film on the heat transfer between the screw and compressed air is assumed to be negligible in this study.

The five steady heat-transfer boundary conditions are described by Eq. (3).

$$k\frac{\partial T}{\partial n} = q_a, \quad a = 1, ..., 5$$

(3)

The parameter q_a represents the five steady heat fluxes at the five different boundaries.

Solving Eqs. (1), (2) and (3) by the finite element/difference method typically requires the assignment of the initial temperature. The calculation continues until the difference between the time-dependent temperature in a certain cycle and that in the previous cycle is small enough. Then, the converged, transient temperature in the rotor during a cycle is obtained. In doing so, the calculation time is usually long and may be affected by the initial condition of temperature. To circumvent the above-mentioned shortcomings, the auxiliary complex equation (ACE) method [17], which gives the steady-state solution of heat conduction problem subject to a periodic boundary condition, is used in this study. Equations (4), (5), (6), (7) and (8)are derived from Eqs. (1), (2) and (3), which describe a boundary-value problem involving a periodic boundary condition. The calculation time required to solve Eqs. (4), (5), (6), (7) and (8) is much shorter than that required to solve Eqs. (1), (2) and (3).

$$\nabla^2 \psi_m - \frac{i \cdot \omega_m}{\alpha} \psi_m = 0$$

(4)

$$k\frac{\partial \psi_m}{\partial n} = h \cdot (A_{m,g} - \psi_m)$$

(5)

$$k\frac{\partial \psi_m}{\partial n} = q_a, \quad a = 1,....5, \quad \text{when } m \neq 0, q_a = 0$$

(6)

$$T_d = \sum_{m=0}^{\infty} \psi_m \cdot \exp(i \cdot \omega_m \cdot \tau)$$
$$\text{where } \tau = \text{rem}(t, t_p)$$

(7)

$$T = Re(T_d)$$

(8)

where ψm is the space function, T_d is the complex temperature, T is the temperature, and ω_m and $A_{m,g}$ are, respectively, the frequency and the amplitude of the half-range cosine expansion of the time-dependent temperature of the compressed air (T_g) [25].

There are three advantages of calculating the temperature in the rotor using the ACE method described above. The first is that the solution is independent of the initial condition of temperature. The second is that the original time-dependent three-dimensional system [Eqs. (1), (2) and (3)] is converted to a time-independent three-dimensional system [Eqs. (4), (5) and (6)]. This significantly reduces the computation time. The third is that the temperature at any time during a cycle can be efficiently obtained by Eqs.(7) and (8) without the use of interpolation, which usually increases the calculation time.

Boundary Conditions

The boundary conditions of the mathematical model are discussed below. Equation (5) is the periodic boundary condition that describes the heat convection between the screw and the compressed air. In this study, the time-dependent properties of the compressed air and the performance of the oil-injected screw compressor are calculated by a lumped parameter analysis [19]. The errors of the calculated volumetric and isentropic efficiencies are both less than 2%. The error of the calculated pressure curve is less than 5%. Fig. 2 shows the calculated and the measured pressure curves of the compressed air under different operating conditions.

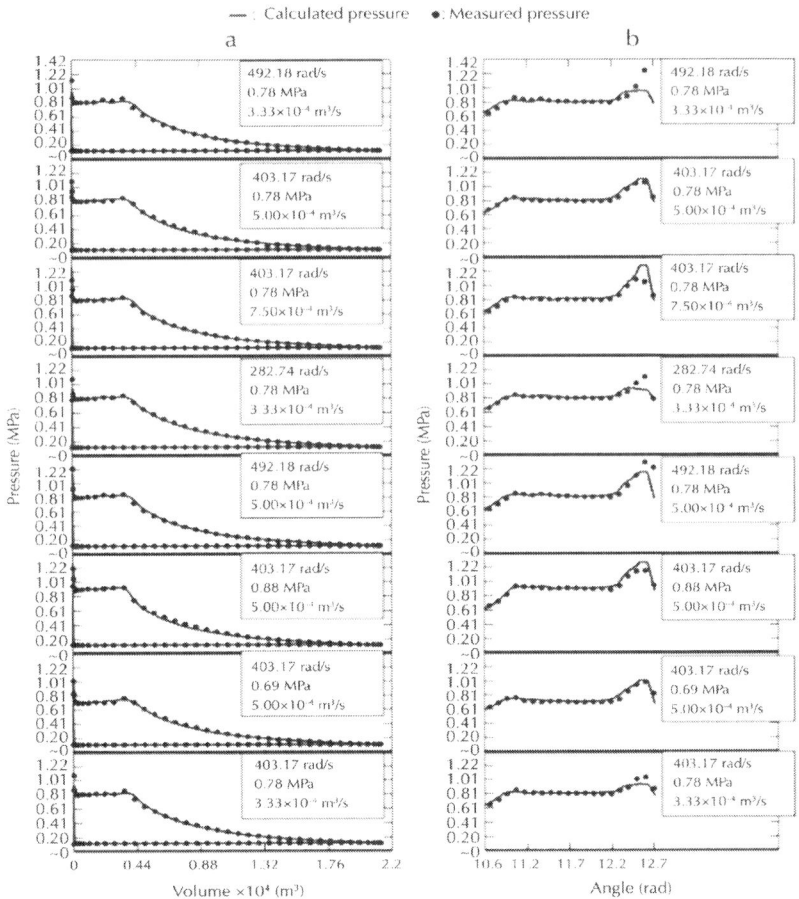

Figure 2: Comparison of the calculated and measured pressures of the compressed air under different operating conditions. The pressure of the compressed air in the screw compressor is calculated by a lumped parameter analysis [19]. The figure shows (a) the PV diagrams of the compressed air during a compression cycle and (b) the pressure curves during the discharge process. The errors of the calculated pressure curves are less than 5%.

In Eq. (5), the convective heat-transfer coefficient h is calculated as the product of an empirical constant ξ_g and an ideal convective heat-transfer coefficient h_{ideal}. The empirical constant ξ_g represents the ratio of the actual to the ideal convective heat-transfer coefficients, and is determined by an optimization procedure to be discussed later in this paper. The empirical correlation, proposed by Gnielinski [18], for the internal forced convection in a pipe is used to calculate h_{ideal}, as shown in Eqs. (9) and (10). The actual convective heat-transfer coefficient between the screw and the compressed air is calculated by Eq. (11).

$$Nu_D = \frac{(f/8)(Re_D - 1000)Pr}{1 + 12.7(f/8)^{1/2}(Pr^{2/3} - 1)}\left[1 + \left(\frac{D_H}{L}\right)^{2/3}\right]$$

$$5000 < Re_D < 10^8, 100 < \frac{D_H}{\varepsilon} < 10^6 \tag{9}$$

$$h_{ideal} = \frac{Nu_D \cdot k}{D_H} \tag{10}$$

$$h = \xi_g \cdot h_{ideal} \tag{11}$$

where D_H is the hydraulic diameter of the compression chamber, and L is the forward displacement of the cross section of the screw, which is calculated by multiplying the lead of screw and the angle of 2ϖ rad s. The friction factor f is calculated by Eq. (12) [26].

$$f = \frac{0.25}{\left[\log\left(\frac{1}{3.7(D_H/\varepsilon)} + \frac{5.74}{Re_D^{0.9}}\right)\right]^2} \tag{12}$$

where ε is the roughness.

The five steady boundary conditions (heat fluxes at five boundaries) are determined in a manner similar to that of the convective heat-transfer coefficient between the screw and the compressed air, as described above. Each of the five steady heat fluxes is calculated as the product of an empirical constant and an ideal heat flux. The empirical constant represents the ratio of the actual to the ideal heat fluxes, and is determined by an optimization procedure. The ideal heat fluxes from the oil to the screw at the sealing line, the inlet end of the screw and the outlet end of the screw are calculated with the formulas given by Xing[22]. The ideal heat flux at the sealing line is calculated by Eq. (13) and the actual heat flux is calculated by Eq. (14).

$$q_{of,\text{ideal}} = \frac{\mu}{4\delta_f}\omega^2 D_o^2$$

(13)

$$q_{of} = \xi_{of} \cdot q_{of,\text{ideal}}$$

(14)

where μ is the dynamic viscosity of oil, δ_f is the clearance of the sealing line, ω is the rotational speed, and D_o is the outer diameter of the screw. ξ_{of} represents the ratio of the actual to the ideal heat fluxes form the oil to the screw at the sealing line.

The ideal heat fluxes from the oil to the screw at the inlet and outlet ends of the screw are respectively calculated by Eqs. (15) and (17). The actual heat fluxes are calculated by Eqs. (16) and (18).

$$q_{os,\text{ideal}} = \frac{\mu}{\delta_s}\omega^2 r^2$$

(15)

$$q_{os,\text{ideal}} = \xi_{os} \cdot q_{os,\text{ideal}}$$

(16)

$$q_{os,\text{ideal}} = \frac{\mu}{\delta_d}\omega^2 r^2$$

(17)

$$q_{od} = \xi_{od} \cdot q_{od,\text{ideal}}$$

(18)

where δ_s and δ_d are the clearances of the inlet and outlet ends of the screw, respectively, and r is the radial distance. The empirical constants ξ_{os} and ξ_{od} represent the ratios of the actual to the ideal heat fluxes.

The ideal heat flux from a bearing to the rotor shaft is determined by dividing the bearing power loss by the area of the bearing. The bearing power loss is calculated by Eq. (19)[21].

$$Q_{b,}\text{ideal}=0.001\times\omega\times M$$

(19)

where ω is the rotational speed, and M is the summation of the torques exerted on the bearing, as given by Eq. (20).

$$M = \begin{cases} M_l + M_v & \text{for the ball bearing} \\ M_l + M_v + M_f & \text{for the roller bearing} \end{cases}$$

(20)

where M_l is the torque due to applied load, M is the torque due to lubricant viscous friction, and M_f is the torque due to roller end-ring flange sliding friction.

The torque M_l due to applied load is calculated by Eq. (21) [21].

$$M_l = f_l \cdot d_m \cdot F_\gamma$$

$$\text{where } f_l = \begin{cases} 0.00028 & \text{for the ball bearing} \\ 0.0015 & \text{for the roller bearing} \end{cases}$$

(21)

The parameter f_l is a factor depending on the bearing design and relative bearing load. The parameter d_m is the bearing pitch diameter. F depends on the magnitude and direction of the applied load, and is calculated by Eq. (22) for the ball bearing or by Eq. (23) for the roller bearing.

$$F_{\gamma,\text{ball}} = 0.9 F_{\text{axial}} \cot(\beta) - 0.1 F_{\text{radial}}$$

(22)

$$F_{\gamma,\text{roller}} = 0.8 \cdot F_{\text{axial}} \cdot \cot(\beta)$$

(23)

where the parameter β is the contact angle of the bearing. The procedure for calculating the average axial force F_{axial} and the average radial force F_{radial} exerting on a bearing is proposed by Stosic et al. [5] and Lee et al. [13]. In their studies, the effects of the pressure force exerted by the compressed air and the input torque on the rotors as well as the contact forces between the male and female screws were considered.

The torque M due to lubricant viscous friction is calculated by Eq. (24) [20].

$$M_\nu = 2.29 \times 10^{-16} \cdot f_o \cdot (\nu_o \cdot 2\pi \cdot \omega)^2 \cdot d_m^3$$

$$\text{where } f_o = \begin{cases} 6.6 & \text{for the ball bearing} \\ 8 & \text{for the roller bearing} \end{cases}$$

(24)

The parameter f_o is a factor depending on the type of bearing and the lubrication system; ν_o is the kinematic viscosity for the lubricant.

In addition, the torque M_f due to roller end-ring flange sliding friction for a roller bearing is calculated by Eq.(25) [21].

Mf=0.006·Faxial·dm   (25)

The actual conductive heat fluxes from the bearing located at the inlet and outlet shafts of the rotor are respectively calculated by Eqs. (26) and (27).

$$q_{bs} = \xi_{bs} \cdot Q_{bs,ideal}/A_{bs}$$

(26)

$$q_{bd} = \xi_{bd} \cdot Q_{bs,ideal}/A_{bd}$$

(27)

where the empirical constants ξ_{bs} and ξ_{bd} represent ratios of the actual to the ideal heat fluxes, and A_{bs} and A_{bd} are the bearing areas.

EXPERIMENTAL SETUP

To provide experimental data for the determination of empirical constants, an experimental facility [19] is adopted in this study. Fig. 3 shows the schematic diagram of the experimental facility. A DC motor operates the screw compressor. A torque meter and a rotational speed meter are installed on the transmission shaft to measure the input torque and the rotational speed of the shaft, respectively. During operation, the air flows into the compressor through the inlet, and oil is injected into the compressor through an injecting port. Inside the compression chamber of the screw compressor, the air and oil are mixed and compressed. The compressed mixed fluid discharges through the outlet of the screw compressor and flows into an oil/gas separator to separate the air and the oil. The separated air and oil then flow through a cooler through separate pipes. After cooling, the air is dried by a water filter. The dried air passes through a back-pressure regulator and an air flow meter. Then the air is discharged to the ambient environment. On the other hand, the cooled oil flows through an oil flow meter and is re-injected into the compressor.

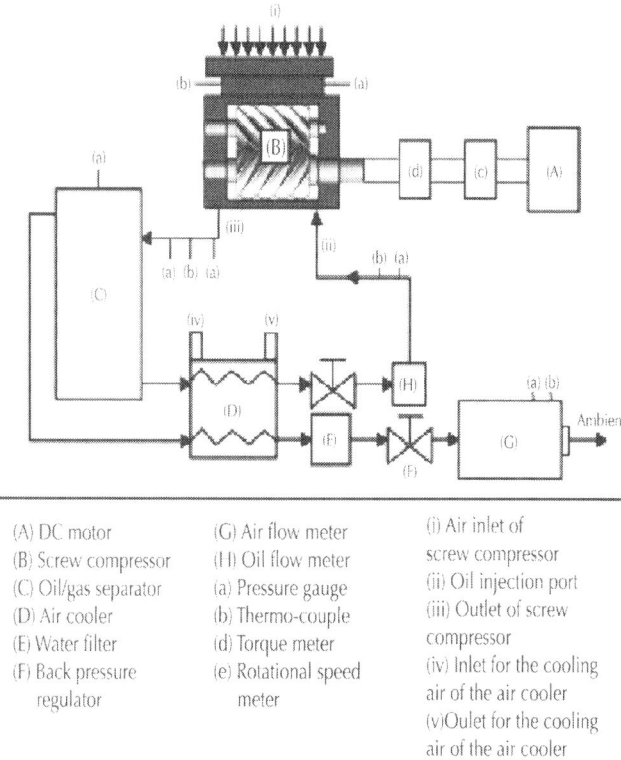

(A) DC motor	(G) Air flow meter	(i) Air inlet of
(B) Screw compressor	(H) Oil flow meter	screw compressor
(C) Oil/gas separator	(a) Pressure gauge	(ii) Oil injection port
(D) Air cooler	(b) Thermo-couple	(iii) Outlet of screw
(E) Water filter	(d) Torque meter	compressor
(F) Back pressure	(e) Rotational speed	(iv) Inlet for the cooling
regulator	meter	air of the air cooler
		(v)Oulet for the cooling
		air of the air cooler

Figure 3: Schematic diagram of the experimental facility [19].

To measure the pressures of both the air and oil, pressure gauges are installed on the outside of the inlet and on the outlet of the compressor, the oil/gas separator and the outlet of the air flow meter. For temperature measurements, thermocouples are installed at the inlet and outlet of the compressor, the oil-injection line and the outlet of the air flow meter. To measure the flow rates of the air and the oil, an air flow meter and an oil flow meter are installed in the system. To determine the PV diagram during a test, the pressure inside the compression chamber is measured by two pressure transducers (Kulite-XTL-140(M)), and the rotational angle is measured by an encoder (HEIDENHAIN ROD 486), as shown in Fig. 4. One pressure transducer is located in the male screw, 3×10^{-3} m from the entrance end of the screw for measuring the pressure during the suction at the beginning of the compression process. The other pressure transducer is installed in the female screw,

3×10^{-3} m from the discharge end of the screw and measures the pressure during both the compression and the discharge process. To measure the temperature distributions in the male and female rotors during a test, thermocouples are installed in the rotors 5×10^{-3} m from the inlet end of the screw, at middle of the screw and 5×10^{-3} m from the outlet end of the screw. The outputs from both the pressure transducers and the thermocouples are recorded, through two slip rings, by a data acquisition A/D card (NI PCI-6071E) installed on a personal computer. The cold-junction temperature of the thermocouples is measured by a resistance temperature detector (Centenary LMPT102). The experimental setup is employed to measure the time-dependent pressure of the compressed air and temperature distributions in the male and female rotors as the screw compressor operates steadily.

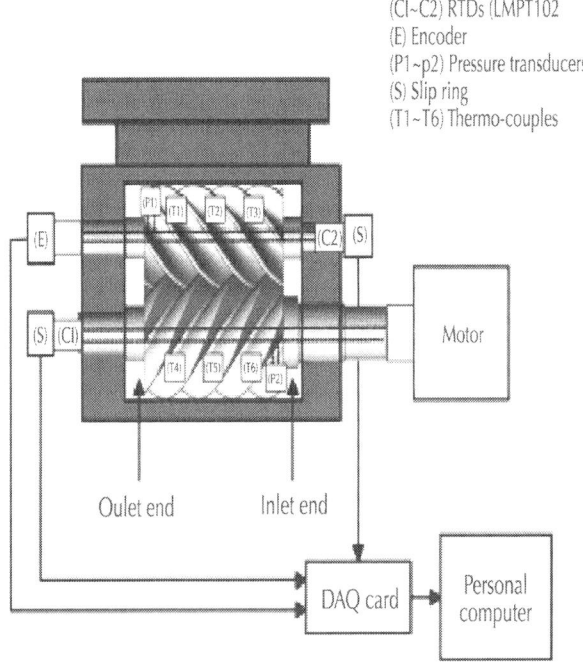

(C1~C2) RTDs (LMPT102
(E) Encoder
(P1~p2) Pressure transducers
(S) Slip ring
(T1~T6) Thermo-couples

Figure 4: Sensors and data acquisition system for the determination of the PV diagram and the temperature of the screws.

Table 2 shows the five test conditions of this study. Three of these five test conditions are adopted to compare the effect of the rotational speed (test conditions 1, 2 and 3) and the discharge pressure (test conditions 1, 4 and 5). The measured results are used to verify the mathematical model and to determine the empirical constants. The uncertainties of the experimental apparatus used in this experiment are specified by the apparatus suppliers. The thermocouple (Omega TT-E-36-72) for measuring the temperature of screw during compression has an uncertainty of ±0.5%. The pressure transducer (Kulite-XTL-140(M)) for measuring the pressure during compression has an uncertainty of ±0.5%. The data acquisition A/D card (NI PCI-6071E), which acquires the data from the thermocouple and the pressure transducer, has an uncertainty of ±0.072%. The resistance temperature detector (Centenary LMPT102), which acquires the temperature of the cold junction inside the screw, has an uncertainty of ±0.08%, and the data acquisition apparatus (CHY504), which acquires the data from the thermal sensor, has an uncertainty of ±0.05%. The temperature meter used by the flow meter has a temperature uncertainty of ±0.38%, and the pressure meter used by the flow meter has an uncertainty of ±0.125%. The uncertainties of the pressure transducer and the pressure meter are calibrated with a pressure gauge (Pennwalt 61A-1A-0050) that has a pressure accuracy of ±0.066%. The uncertainty of the thermocouple and the temperature meter is calibrated with a resistance temperature detector (Omega PR-11) that has a temperature accuracy of ±0.35 K at 373.15 K. The uncertainties of the deduced data are determined with the t-distribution with a confidence level of 0.95 [27] and [28]. The degrees of freedom are determined by the number of test data. The pressure measurement has an uncertainty of ±3.22%. The uncertainty of the temperature in the rotor is ±2.14%. The uncertainty of the volumetric efficiency is about ±0.84%. The uncertainty of the isentropic efficiency is ±0.85%.

Table 2: Test conditions

Experimental condition[a]	Rotational speed (rad/s)	Discharge pressure (MPa)
1	282.74	0.78
2	403.17	0.78
3	492.18	0.78
4	282.74	0.59
5	282.74	0.69

[a]The temperature and flow rate of injected oil are 54 °C and 5 × 10⁻⁴ m³/s, respectively, under all test conditions.

CALCULATION PROCEDURE AND THE GRID INDEPENDENCE TEST

Calculation Procedure

The flow chart given in Fig. 5 shows the four steps in the computing procedure used to solve the above-mentioned mathematical model and to determine the six empirical constants through the minimization of the difference between the calculated and measured temperature of the rotors. The first step is to input the geometric parameters of rotors, the five test conditions, and the measured results under the five test conditions. In the second step, the three-dimensional geometric models of rotors, the meshes and the governing equations are generated. The third step is to calculate the ideal convective heat-transfer coefficient [Eq. (10)], ideal heat fluxes [Eqs. (13), (15) and (17)] and power generated by the bearings [Eq.(19)] from the temperature of compressed air, the frictional power generated by the oil and the power generated by the bearing, under the five test conditions. In the forth step, the mathematical model is solved and the difference between the measured and calculated results under the five sets of operating conditions is minimized by the method of sequential quadratic programming (SQP) [29], [30], [31] and [32] to determine the six empirical constants.

1. Input	2. 3D Solid Models	3. Ideal physical parameters
(1) Geometric parameters of screw compressor (2) J sets of operating conditions (3) J sets of measured data (e)	(1) Create 3D models (2) Mesh (3) Generate J sets of governing equations (GV)	Calculate J sets of ideal (1) convew=ctive heat-transfer coefficient [Eq. (10)] (2) heat fluxes [Eqs. (13)(15)(17)] (3) power generated by bearings [Eq. (19)]

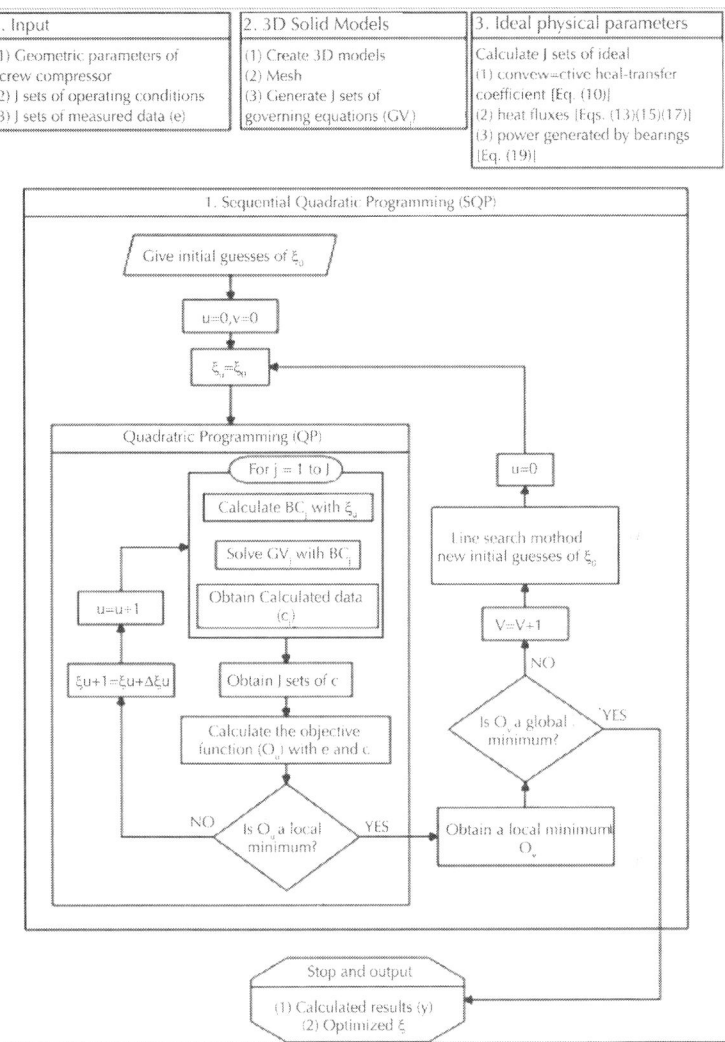

Figure 5: Flow chart showing the four steps in the computing procedure to solve the mathematical model and to determine the six empirical constants through the minimization of the difference between the calculated and measured temperature of the rotors.

During the sequential quadratic programming of optimization, the initial values of the six empirical constants are input to the program. Then, an inner-loop iteration is conducted by using Quadratic Programming to determine a set of the six empirical conditions under a local minimum condition. In the inner-loop, the mathematical model [Eqs. (4), (5), (6), (7) and (8)] under five operating conditions is solved by the commercial program COMSOL Multiphysics. The solution of the mathematical model is compared with the measured data and the value of the objective function, given in Eq. (28) is determined. The program repeatedly calculates the objective function by updating the values of the empirical constants until a local minimum of the objective function is obtained.

$$\text{Objective Function } = \sum_{j=1}^{5} \sum_{\eta=1}^{3} \left| \frac{e_{\eta j} - c_{\eta j}}{e_{\eta j}} \right|$$

(28)

e: measured temperature; c: calculated temperature; η: index of the position; $\eta = 1$: 0.005 m from the inlet end of the screw; $\eta = 2$: middle of the screw; $\eta = 3$: 0.005 m from the outlet end of the screw; j : index of the test condition.

When the inner-loop iterates until the changes in the empirical constants are all less than 10^{-6}, a local minimum of the objective function is obtained. Then, an outer loop of the iteration begins by performing a line search calculation to provide a new set of initial guesses of the empirical constants for the next inner-loop iteration. The next inner-loop iteration is then performed, as described in the last paragraph, by using the Quadratic Programming to determine another set of the six empirical conditions based on the new set of initial guesses of the empirical constants provided by the outer-loop iteration. The outer-loop iteration will repeat the above-mentioned procedures to obtain new local minimums until a global minimum of the objective function is found. The calculation of the SQP (outer-loop) iterates until the changes in the objective function and empirical constants are all less than 10^{-6}.

Grid Independence Test

The meshing tools in the commercial program COMSOL Multiphysics are employed to establish the meshes. A grid independence test is carried out to determine a set of meshes by which accurate results can be obtained efficiently. In the test, there are twelve sets of meshes created by the combination of the four different swept meshes created

by the swept mesher and the three different free meshes generated by
the free mesher of the program. The four swept meshes on the screws
have 8, 18, 28 and 38 layers of elements. The three free meshes are the
Normal Grid, Fine Grid and Finer Grid, as predefined by the meshing
tools. The numbers of elements of the twelve meshes are listed in Table
3.

Table 3: Grid independence test

Number of element layers of swept mesh		8	18	28	38
Number of elements					
Predefined type of free meshes	Normal grid	22714	30414	38132	45892
	Fine grid	27775	37482	47336	57158
	Finer grid	48862	48919	63043	76464
Maximum error, %					
Predefined type of free meshes	Normal grid	0.1685	0.0624	0.0617	0.0696
	Fine grid	0.1537	0.0614	0.0381	0.0451
	Finer grid	0.1570	0.0539	0.0285	0.00[a]

[a]The base-line condition.

Temperature distributions in the male and female rotors are
calculated with the twelve meshes under identical initial and boundary
conditions. The set of temperature distributions calculated with the
mesh that has the most elements (38 element layers and a finer grid
in Table 3) is considered as the base-line condition, and is compared
with the other eleven sets of temperature distributions obtained with
the other meshes. During comparison, the relative errors at 160,000
locations of the rotors are calculated, and the maximum relative errors
are listed in Table 3.

As shown in Table 3, it is evident that the effect of number of the
element layers of the swept meshes on the calculated maximum error
is larger than that of the free meshes. When the number of elements
is larger than 40,000, the maximum error is less than 0.06% and the
calculated temperature distributions are independent of the number of
elements. Based on the results of the grid independence test, the set
of meshes generated with 18 element layers in the swept meshing and
the Finer Grid in the free meshing, as shown in Fig. 6, is used in the
calculation of temperature distributions.

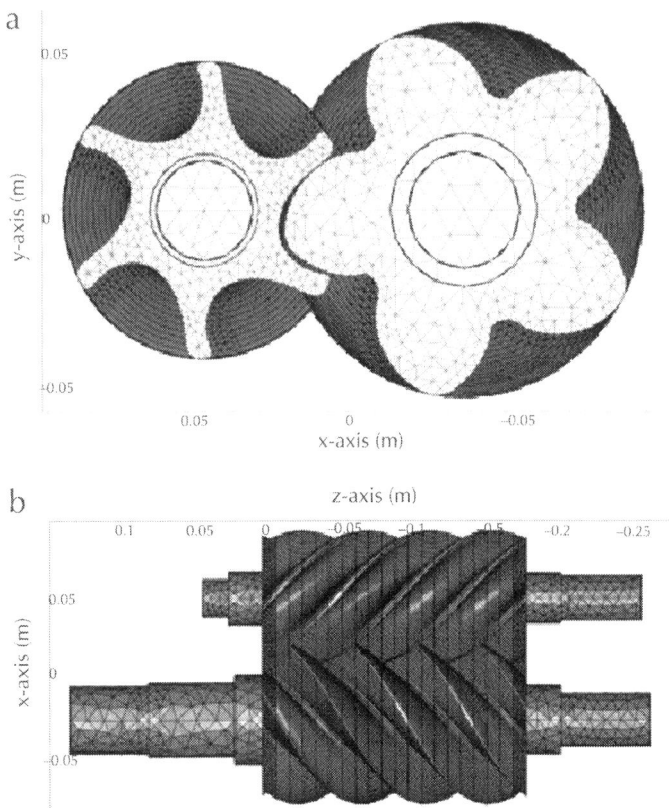

Figure 6: The meshes are generated with 18 element layers in the swept meshing and the Finer Grid in the free meshing. There are 33,098 elements in the male rotor and 15,821 elements in the female rotor.

RESULTS AND DISCUSSION

In this study, the numerical method described in the above section is adopted to solve mathematical model and to determine the six empirical constants through the minimization of the difference between the calculated and measured temperatures of the rotors under the five sets of operating conditions listed inTable 2. Fig. 7 shows the comparison of the calculated and measured temperatures of the rotors under the five operating conditions listed in Table 2. The average errors in the calculated temperature at three locations of each of the male and female rotors are 5.45% and 4.85%, respectively. The average error is calculated by Eq. (29).

$$\text{Average error} = \left(\sum_{j=1}^{5} \sum_{\eta=1}^{3} \left| \frac{e_{\eta j} - c_{\eta j}}{e_{\eta j}} \right| \right) \Big/ 15$$

(29)

e: measured temperature; c: calculated temperature; η: index of the position; $\eta = 1 : 0.005$ m from the inlet end of the screw; $\eta = 2 :$ middle of the screw; $\eta = 3 : 0.005$ m from the outlet end of the screw; j : index of the test condition

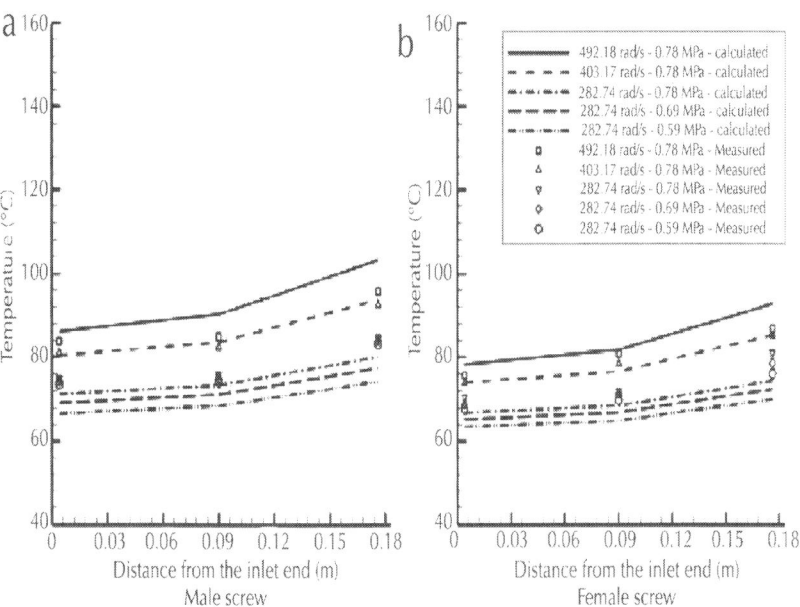

Figure 7: Comparison of the calculated and measured temperatures of the rotors under the five operating conditions listed in Table 2. The average errors of the calculated temperatures at three locations of each of the male and female rotors are 5.45% and 4.85%, respectively.

Both the calculated and experimental results indicate four characteristics of temperature distributions in rotors. The first characteristic is that, under the same operating conditions, the temperature in the male rotor is greater than that in the female rotor at the same distance from the inlet end of the screw. This is caused by the fact that bearings installed on the male rotor are subjected to larger frictional forces and induce more power losses than the ones installed on the female rotor (see Table 4, which shows the calculated bearing forces and power losses under the five test conditions listed in Table 2). This causes more heat is dissipated form the bearings installed on the male rotor and induces a higher temperature in the male rotor. The second characteristic is that the temperature in the rotor increases with increasing distance from the inlet end of the screw. This behavior occurs because the heat is mainly transferred from the bearings installed on the outlet shaft of the rotor to the screw. Then, the heat is transferred from the screws to both the compressed air and the bearings installed on the inlet shaft of the rotors. The third characteristic is that the temperature increase from the middle to outlet end of the screw is larger than that from the inlet end of the screw to the middle. This occurs because the compressed air takes away part of the heat inside the screw via convective heat transfer. The heat conduction along the screw is reduced at positions closer to the inlet end of the screw. Therefore, the temperature gradient in the axial direction is different at different positions along the screw. The last characteristic is that the temperature increases with increasing rotational speed and discharge pressure. This effect occurs because higher rotational speeds and discharge pressures lead to larger friction power losses in the bearing. The temperature of the rotor, which is mainly affected by the friction power loss of the bearing, increases with increasing rotational speed and discharge pressure.

Table 4: The bearing force intensities and power losses

Test condition listed in Table 2	Axial force (N)	Radial force (N)		Power loss (W)	
		Inlet shaft	Outlet shaft	Inlet shaft	Outlet shaft
The bearings on the shafts of the male screw					
1	3404.42	754.51	1701.21	23.33	87.36
2	3460.84	776.80	1721.60	35.60	129.07
3	3486.11	792.74	1748.10	45.91	161.87
4	2874.40	705.05	1488.18	21.87	71.96
5	3166.57	744.41	1591.35	23.03	80.14

The bearings on the shafts of the female screw					
1	716.65	990.31	1962.37	3.70	19.25
2	725.85	1018.51	1990.89	5.57	28.29
3	724.52	1038.69	2016.26	7.10	35.41
4	571.27	917.78	1696.32	3.43	16.33
5	647.75	972.22	1835.64	3.63	17.85

The Temperature Distribution in Rotors

The temperature distributions in the male and female rotors, which operated steadily under a rotational speed of 403.171 rad/s, outlet pressure of 0.784 MPa and oil-injected velocity of 5×10^{-4} m³/s, were calculated by the verified mathematical model. The calculated results show that during an operating period both of the local largest changes of temperature in the male and female rotors occur at the surfaces of the screws near the outlet ends. The temperature changes are smaller for positions closer to the inlet end of the screws. The largest local change of temperature in the male rotor is 0.0156 °C. The largest local change of temperature in the female rotor is 0.0071 °C. These results show that the changes of temperature in the male and female rotors, which operated steadily, are small during an operating period.

Fig. 8(a) shows the six locations of each of the male and female rotors for the temperature distributions given in Fig. 8(b). (P_A, P_G: the centers, P_B, P_H: the load carrying flanks, P_C, P_I: the top lands, P_D, P_J: the back flanks, P_E, P_K: the bottom lands and P_F, P_L: the mass centers of the teeth) The twelve temperature curves indicate two features of temperature distributions in the male and female rotors, which operate steadily. The first feature is that the temperature in the male rotor is larger than that in the female rotor at the same distance from the inlet end. The reason is that bearings installed on the male rotor are subjected to larger frictional forces and induce more power losses than the ones installed on the female rotor. This causes more heat is dissipated form the bearings installed on the male rotor and induces a higher temperature in the male rotor.

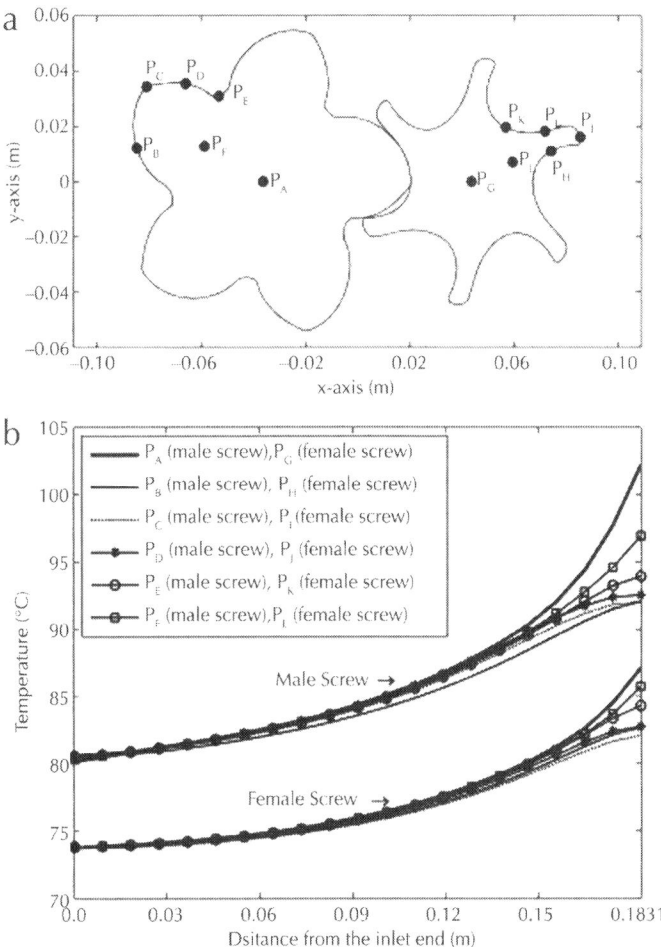

Figure 8: (a) Schematic diagram showing the six locations of each of the male and female rotors for the temperature distributions in (b). P_A, P_G: the centers, P_B, P_H: the load carrying flanks, P_C, P_I: the top lands, P_D, P_J: the back flanks, P_E, P_K: the bottom lands and P_F, P_L: the mass centers of the teeth. (b) The calculated temperature distributions at the six locations of each of the male and female rotors.

The second feature is that the temperature in the rotor increases with increasing distance from the inlet end of the screw. The temperature increase from the middle (0.0916 m from the inlet end of the screw) to outlet end of a screw (0.1831 m from the inlet end of the screw) is larger than that from the inlet end to the middle. Larger gradient of temperature along the axial direction implies larger rate of heat conduction along the axial direction. This feature shows that the heat is mainly transferred from the bearings installed on the outlet shaft of the rotors into the screws. Then, the heat is transferred from the screws to the compressed air and the bearings installed on the inlet shaft of the rotors. The compressed air takes away part of the heat inside the screw via convective heat transfer. The heat conduction along the screw is reduced at positions closer to the inlet end of the screw. Therefore, the temperature gradient in the axial direction is different at the different position along the screw.

Fig. 9 shows the isotherm graphs at the three cross sections of the male and female rotors [(a) 0.0092 m from the inlet end, (b) 0.0916 m from the inlet end and (c) 0.1739 m from the inlet end]. Near the inlet end of the screw, the temperature distributes uniformly in the male and female screws, as shown in Fig. 9(a). The maximum temperature differences at this cross section are 0.24 °C and 0.10 °C for the male and female screws, respectively. In the middle of the screw, Fig. 9(b) shows that the temperature distributes uniformly in the regions bounded by the root circles of the male and female screws. Lower temperatures exit in the area near the back flank of the male screw and in the area between the back flank and the top land of the female screw. The local maximum temperature differences on the cross sections are 0.82 °C and 0.52 °C for the male and female screws, respectively. Fig. 9(c) shows the temperature distribution on the cross sections close to the outlet end of the screw. It is noted that the temperature at the center is the highest in the cross section of each of the male and female screws. This indicates that heat is transferred radially from the center to the surface of the screw and is taken away by the force convection of the compressed air at the surface of the screw. The temperature differences between the centers and the surfaces of the male and female screws are 6.36 °C and 2.98 °C, respectively.

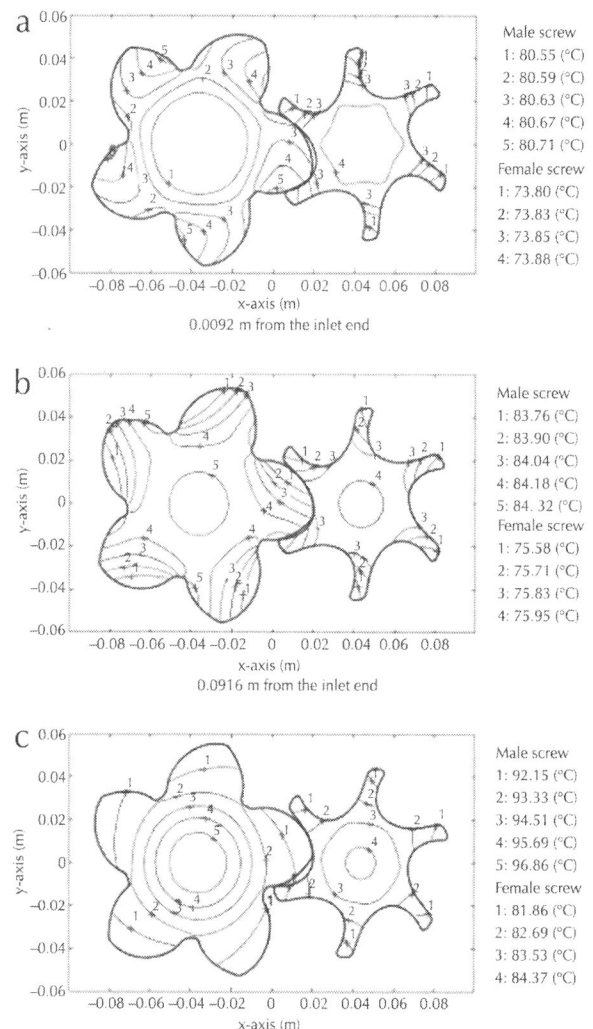

Figure 9: Isotherm graphs in the three cross sections of the male and female rotors. (a) 0.0092 m from the inlet end, (b) 0.0916 m from the inlet end and (c) 0.1739 m from the inlet end.

A comparison of Fig. 9(a) with Fig. 9(b and c) reveals that the temperature difference on the cross section near the outlet end of the screw is larger than those on the other two cross sections. This shows that the heat transferred radially from the center to the surface of the screw is larger at the section near the outlet end than that at the other two cross sections that are located at the middle and inlet end. In addition, the temperature gradient in the radial direction of the screw near the outlet end is larger than the one near the inlet end. One of the reasons would be that the temperature increase gradient of the oil and air mixture inside the compression chamber near the outlet end is larger than that near the inlet end, as shown inFig. 10.

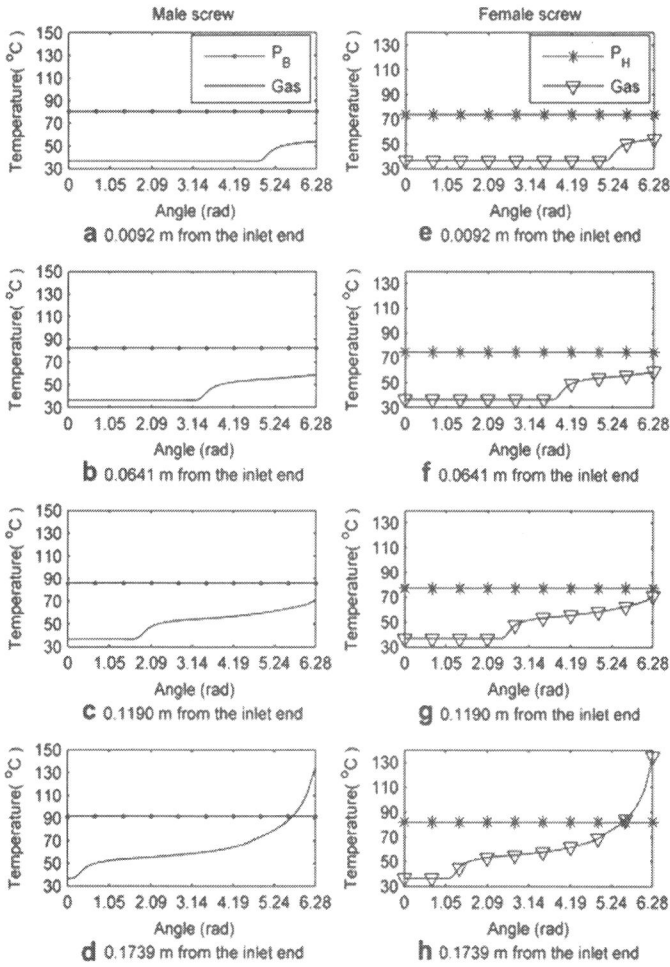

Figure 10: Calculated temporal temperature curves of the screw at the load carrying flanks (P_B, P_H) and the compressed air on the four cross sections of the male [(a) – (d)] and female rotors [(e) – (h)] during an operating cycle.

Fig. 10 depicts the calculated temporal temperature curves of the screw at the load carrying flanks (P_B, P_H) and the compressed air, on the four cross sections of the male [(a)–(d)] and female rotors [(e)–(h)], during an operating cycle. The four cross sections are located respectively at 0.0092 m, 0.0641 m, 0.1190 m and 0.1739 m from the inlet end of the screw. Fig. 10 shows that the temperatures at the load carrying flanks (P_B, P_H) of the male and female screws are always higher than those of the compressed air on the cross sections at 0.0092 m, 0.0641 m and 0.1190 m from the inlet end of the screw. This condition reverses only when the compression angle is larger than about 5.59 rad on the cross sections at 0.1739 m from the inlet end of the screw. This indicates that the heat is transferred from the screw to the compressed air at most parts of the screws and during most of the compression cycle. The convective heat-transfer, however, is significantly larger about ten times larger when the condition reverses (the compression angle is larger than 11.17 rad on the cross sections positioned at 0.1739 m from the inlet end of the screw) due to the high temperature and density of the compressed air during the discharge process. In Fig. 10, it is evident that the temperature increase gradient of the oil and air mixture inside the compression chamber near the outlet end is larger than that near the inlet end.

The information of the temperature distributions in the screws can be applied to estimate the approximate thermal deformation of the screws. The coefficient of thermal expansion of the rotors adopted in this study is 1.08×10^{-5} m/(m °C). The average temperature rise of the male screw is about 62.3 °C, under test condition 2 of Table 2, and the induced axial thermal expansion and radial thermal expansion are about 1.23×10^{-4} m and 3.80×10^{-5} m, respectively. On the other hand, the average temperature rise of the female screw is about 56.2 °C, the axial thermal expansion is about 1.11×10^{-4} m, and the radial thermal expansion is about 2.72×10^{-5} m. For the male rotor, the average temperature rise on the outlet end of the screw is about 11.4 °C higher than that on the inlet end of the screw; the temperature is 11.3 °C higher at the outlet end than that at the inlet end for the female rotor. This higher temperature rise induces higher thermal deformation and stresses in the screw near the outlet end than at the inlet end.

Sensitivity Analysis of the Six Empirical Constants

A global sensitivity analysis (GSA) [33], [34] and [35] is conducted to study the effects of the six empirical constants that are used to calculate the six heat-transfer boundary conditions on the temperature distributions in the rotors, and to determine how many empirical constants are needed for the mathematical model to accurately predict the temperature distributions in the rotors. First, 240 sets of empirical constants were randomly selected by the Monte Carlo Method. The minimum values of the six empirical constants in the sensitivity analysis are zeros, which means there is no heat fluxes on the boundary surfaces. The maximum values of the six empirical constants in the sensitivity analysis are ones. This implies that all the energy generated under ideal conditions is transferred to the rotors.

Next, the authors calculate 240 sets of temperature distributions of the male and female rotors using the 240 random sets of empirical constants. Four outputs of the mathematical model are determined from the calculated 240 sets of temperature distributions. For both of the male and female screws, the four outputs of the mathematical model are the average temperature of the screw and the axial temperature gradients in the inlet part (0.0–0.061 m from the inlet end of the screw), the middle part (0.061–0.122 m from the inlet end of the screw) and the outlet part (0.122–0.183 m from the inlet end of the screw) of the screw. The three axial temperature gradients are taken to be the slopes of the first-order linear equations calculated by fitting the temperatures on the meshes.

In the sensitivity analysis, the authors adopted the method introduced by Andrea Saltelli [35] that reduces the computing time for obtaining the Sobol's Global Sensitivity Indices to calculate the total sensitivity indices (TSI) of the six empirical constants from the four sensitivity indices of the temperature distributions. The results of the sensitivity analysis are listed in the Table 5. If an empirical constant has a TSI with a large absolute value, it significantly affects the calculated results. In this work, a high TSI also means that the boundary condition calculated by the certain empirical constant has a strong effect on the calculated temperature distributions.

Table 5: The sensitivity analysis of the empirical constants for the temperature distributions in the rotors

| | Empirical constant | Average temperature | Temperature increasing rate | | |
			Inlet part of screw	Middle part of screw	Outlet part of screw
TSI for the male rotor	ξ_{bs}	0.001	0.263	0.522	0.054
	ξ_{bd}	0.666	−0.120	0.594	1.838
	ξ_g	0.734	−0.071	0.003	−0.138
	ξ_{os}	−0.025	−0.035	0.000	−0.144
	ξ_{od}	−0.021	−0.050	−0.008	−0.117
	ξ_{of}	−0.027	−0.049	−0.012	−0.148
TSI for the female rotor	ξ_{bs}	−0.149	0.416	0.324	−0.190
	ξ_{bd}	0.713	−0.030	0.793	1.904
	ξ_g	0.590	−0.014	0.411	0.084
	ξ_{os}	−0.166	0.010	−0.121	−0.287
	ξ_{od}	−0.162	−0.013	−0.130	−0.270
	ξ_{of}	−0.163	0.010	−0.139	−0.292

ξ_{bs}: calculates the heat flux at the inlet shaft of the rotor. Eq. (26).

ξ_{bd}: calculates the heat flux at the outlet shaft of the rotor. Eq. (27).

ξ_g: calculates the heat-transfer coefficient for the heat convection. Eq. (11).

ξ_{os}: calculates the heat flux at the inlet end of the screw. Eq. (16).

ξ_{od}: calculates the heat flux at the outlet end of the screw. Eq. (18).

ξ_{of}: calculates the heat flux at the sealing line of the screw. Eq. (14).

For the first sensitivity index, the average temperature, the empirical constants ξ_g, (for calculating the heat convection between the screw and the compressed air) and ξ_{bd}, (for calculating the heat transfer between the rotor and the bearing at the outlet shaft of the rotor) have the two largest absolute values of the TSI for the two rotors. For the axial temperature gradient in the inlet part of the screw, the empirical constant ξ_{bs} (for calculating the heat transfer between the rotor and the bearing at the inlet shaft of the rotor) has the largest absolute value of the TSI for both the male and female rotors. For the axial temperature gradient in the middle part of the male screw, the empirical constant ξ_{bd}, has the largest absolute value of the TSI, and the empirical constant ξ_{bs}, has the second largest absolute value of the TSI. For the female screw, the empirical constant ξ_{bd}, has the largest absolute value of the

TSI, the empirical constant ξ_g, has the second largest absolute value of the TSI, and the empirical constant ξ_{bs}, has the third largest absolute value of the TSI. For the axial temperature gradient in the outlet part of the screw, the empirical constant ξ_{bd}, has the largest absolute value of the TSI for both the male and female rotors. These findings indicate that the three empirical constants ξ_g, ξ_{bd} and ξ_{bs}, have a stronger impact on the outputs of the mathematical model than the other three empirical constants.

To determine if a mathematical model with less empirical constants is able to accurately predict the temperature distributions in the rotors, the authors performed the following analysis based on the results of the above sensitivity analysis. The temperature distributions were calculated with a reduced mathematical model that uses only the three empirical constants that have the strongest impact on the results. In the reduced mathematical model, the values of the three empirical constants ξ_g, ξ_{bd} and ξ_{bs}, are determined by minimizing the difference between the calculated and measured data (as shown in Fig. 7), and the values of the other three empirical constants are set equal to one.

The determined empirical constants of the original mathematical model, of which the calculated results are shown in Fig. 7, and the reduced mathematical model are listed in Table 6. The difference between the determined empirical constants of the original and the reduced mathematical models are also listed in Table 6. In Table 6, the empirical constant of ξ_{bd} for the female rotor is shown to have the maximum difference of 1.83%. Table 6 also lists the maximum relative difference between the calculated temperature distributions using the original and reduced mathematical models. The maximum relative differences for the male and female rotors are 0.43% and 0.33%, respectively, and are located on the ends of the outlet shafts of the male and female rotors. The calculated results obtained by the reduced mathematical model are similar to those calculated by the original mathematical model. Therefore, the reduced mathematical model, which uses only three empirical constants, can be used to study the temperature distribution in the rotor.

Table 6: The optimization results of the original and the reduced mathematical models

Empirical constant	Original model (6 empirical constants)		Reduced model (3 empirical constants)		The difference between the original and the reduced models, %	
	Male rotor	Female rotor	Male rotor	Female rotor	Male rotor	Female rotor
ξ_{bs}	−0.1955	−0.2997	−0.1990	−0.3054	0.35	0.57
ξ_{bd}	0.5050	0.6952	0.5018	0.7135	0.32	1.83
ξ_{g}	0.3031	0.1145	0.3101	0.1248	0.70	1.03
ξ_{os}	0.5722	−0.2208	1.0[a]	1.0[a]	_[b]	_[b]
ξ_{od}	0.4289	0.1010	1.0[a]	1.0[a]	_[b]	_[b]
ξ_{of}	0.3490	−0.2583	1.0[a]	1.0[a]	_[b]	_[b]
Maximum relative difference[c], %			0.43	0.33		

Maximum relative difference =max $(|T_{orginal;\ \varepsilon_j} T_{reduced;\ \varepsilon_j}|=T_{orginal;\ \varepsilon_j})$; where ε is the index of the grid point and j is the index of the test condition.

[a]The empirical constant is not determined by the optimization, and is set to be one.

[b]The value is null for the empirical constant in the reduced model and is not determined by the optimization.

[c]This value is the maximum relative difference between the calculated temperature on the grid points using the original and reduced mathematical models under the five test conditions listed in Table 2, as defined below. The total number of grid points are 651,384 and 348,912 for the male and female rotors, respectively.

SUMMARY AND CONCLUSIONS

A mathematical model and a calculation procedure are proposed in this study to efficiently calculate the temperature distributions in the male and female rotors of the oil-injected screw compressor. The solution of the transient heat conduction problem of the rotors, which is subject to a periodic boundary condition and five steady boundary conditions, is obtained by solving the set of Helmholtz equations that are derived from the partial differential equations for transient heat conduction without internal heat. During the solving process, the periodic convective and the five steady boundary conditions are calculated with the six empirical constants, the ideal convective heat-transfer coefficient and the ideal steady heat fluxes. The six empirical constants are determined by minimizing the difference between the calculated and measured temperature of the rotors. The average errors of calculated temperature at three locations of each of the male and female rotors are 5.45% and 4.85%, respectively. The main results of this study are as follows.

- Under the same operating conditions, the temperature in the male rotor is larger than that in the female rotor at the same distance from the inlet end of the screw. This is caused by the fact that bearings installed on the male rotor are subjected to larger frictional forces and induce more power losses than the ones installed on the female rotor. This causes more heat is dissipated form the bearings installed on the male rotor and induces a higher temperature in the male rotor.

- The heat is mainly transferred from the bearings, installed on the outlet shaft of the rotors, into the screws. Then, the heat is transferred from the screws to the compressed air and the bearings, installed on the inlet shaft of the rotors. The compressed air takes away part of the heat inside the screw via convective heat transfer. The heat conduction along the screw is reduced at positions closer to the inlet end of the screw. Therefore, the temperature gradient in the axial direction is different at different positions along the screw.

- For the convective heat transfer between the screw and the compressed air, the heat is transferred from the screw to the compressed air at most parts of the screws and most of the time during a compression cycle. The radial heat transfer is larger at sections closer to the outlet end of the screw.

- The three empirical constants, which are used to calculate the heat convection between the screw and the compressed air, the heat transfer between the rotor and the bearing at the inlet shaft of the rotor and the heat transfer between the rotor and the bearing at the outlet shaft of the rotor, have a stronger impact on the outputs of the mathematical model than the other three empirical constants. The reduced mathematical model, which uses only the three empirical constants with the strongest impact, can be used in the studies of the temperature distribution in the rotor.

• The calculated temperature distributions in the screws can be used to estimate the approximate thermal deformation of the screws and thereby improve the screw profiles and the design of the oil-injected screw compressor.

ACKNOWLEDGEMENTS

This paper is supported in part by the Ministry of Economic Affairs, R.O.C. under Contract 94-EC-17-A-05-I1-0006, and was technically supported by Fu Sheng Industrial Co., Ltd.

REFERENCES

1.   N. Stosic, I.K. Smith, A. Kovacevic, Opportunities for innovation with screw compressors, Proceedings of the Institution of Mechanical Engineers, Part E: Journal of Process Mechanical Engineering 217 (2003) 157e170.

2.   K. Hanjalic, N. Stosic, Development and optimization of screw machines with a simulation model 2. Thermodynamic performance simulation and design optimization, Journal of Fluids Engineering e Transactions of the ASME 119 (1997) 664e670.

3.   N. Stosic, K. Hanjalic, Development and optimization of screw machines with a simulation model 1. Profile generation, Journal of Fluids Engineering e Transactions of the ASME 119 (1997) 659e663.

4.   N. Stosic, I.K. Smith, A. Kovacevic, Optimisation of screw compressors, Applied Thermal Engineering 23 (2003) 1177e1195.

5.   N. Stosic, I.K. Smith, A. Kovacevic, Screw Compressors e Mathematical Modelling and Performance Calculation. Springer Verlag Berlin Heidelberg, New York, 2005.

6.   N. Stosic, L. Milutinovic, K. Hanjalic, A. Kovacevic, Investigation of the influence of oil injection upon the screw compressor working process, International Journal of Refrigeration 15 (1992) 206e220.

7.   M. Fujiwara, Y. Osada, Performance analysis of an oil-injected screw compressor and its application, International Journal of Refrigeration 18 (1995) 220e227.

8.   J.S. Fleming, Y. Tang, The analysis of leakage in a twin-screw compressor and its application to performance improvement, Proceedings of the Institution of Mechanical Engineers, Part E: Journal of Process Mechanical Engineering 209 (1995) 125e136.

9.   J.S. Fleming, Y. Tang, G. Cook, The twin helical screw compressor part 1: development, applications and competitive position, Proceedings of the Institution of Mechanical Engineers Part C e Journal of Mechanical Engineering Science 212 (1998) 355e367.

10.   J.S. Fleming, Y. Tang, G. Cook, The twin helical screw compressor part 2: a mathematical model of the working process, Proceedings of the Institution of Mechanical Engineers Part C e Journal of Mechanical Engineering Science 212 (1998) 369e380.

11.   H.G. Wu, Z.W. Xing, P.C. Shu, Theoretical and experimental study on indicator diagram of twin screw refrigeration compressor, International Journal of Refrigeration 27 (2004) 331e338.

12.   N. Seshaiah, S.K. Ghosh, R.K. Sahoo, S.K. Sarangi, Mathematical modeling of the working cycle of oil injected rotary twin screw compressor, Applied Thermal Engineering 27 (2007) 145e155.

13.   W.S. Lee, R.H. Ma, W.F. Wu, S.L. Chen, Performance and bearing load analysis of a twin screw air compressor, Chinese Journal of Mechanics e Series A 15 (1999) 69e78.

14.   A. Kovacevic, Boundary adaptation in grid generation for CFD analysis of screw compressors, International Journal for Numerical Methods in Engineering 64 (2005) 401e426.

15.   A. Kovacevic, N. Stosic, E. Mujic, I.K. Smith, CFD integrated design of screw compressors, Engineering Applications of Computational Fluid Mechanics 1 (2007) 96e108.

16.   [A. Kovacevic, N. Stosic, I.K. Smith, Screw Compressors e Three Dimensional Computational Fluid Dynamics and Solid Fluid Interaction. Springer Verlag Berlin Heidelberg, New York, 2007.

17.   M.N. Ozisik, Boundary Value Problems of Heat Conduction. Dover Publications, New York, 1989, pp. 110e114.

18.   V. Gnielinski, New equations for heat and mass-transfer in turbulent pipe and channel flow, International Chemical Engineering 16 (1976) 359e368.

19.   S.H. Hsieh, Y.C. Shih, W.H. Hsieh, F.Y. Lin, M.J. Tsai, Performance analysis of screw compressors e numerical simulation and experimental verification, Proceedings of the Institution of Mechanical Engineers Part C-Journal of Mechanical Engineering Science submitted for publication.

20.   Y.R. Jeng, P.Y. Huang, Predictions of temperature rise for ball bearings, Tribology Transactions 46 (2003) 49e56.

21.   A.H. Tedric, N.K. Michael, Essential Concepts of Bearing Technology, fifth ed. CRC Press, 2006

22.   Z.W. Xing, Screw Compressors e Theory, Design and Application. China Machine Press, Beijing, 2000.

23.   Y.R. Wu, Z.H. Fong, Rotor profile design for the twin-screw compressor based on the normal-rack generation method, Journal of Mechanical Design 130 (2008) 8.

24.   D.H. Kirk, Heat Transfer with Application, International ed. Prentice Hall, 1998.

25.   D.G. Michael, Advanced Engineering Mathematics, second ed. Prentice Hall, 1998.

26.   P.K. Swamee, A.K. Jain, Explicit equations for pipe-flow problems, Journal of the Hydraulics Division (ASCE) 102 (1976) 657e664.

27.   R.P. Benediet, Fundamentals of Temperature, Pressure and Flow Measurements. John Wiley & Sons, Hoboken, USA, 1984.

28.   S.J. Kline, F.A. McClintock, Describing the uncertainties in single sample experiments, Mechanical Engineering 75 (1953) 3e8.

29.   R. Fletcher, Practical Methods of Optimization. John Wiley and Sons, Hoboken, USA, 1987.

30.   P.E. Gill, W. Murray, M.H. Wright, Practical Optimization. Academic Press, London, UK, 1981.

31.   M.J.D. Powell, Variable Metric Methods for Constrained Optimization. Springer Verlag, 1983.

32.   J.S. Arora, Introduction to Optimum Design, International ed. McGraw-Hill Book Company, 1989.

33. A. Kiparissides, S.S. Kucherenko, A. Mantalaris, E.N. Pistikopoulos, Global sensitivity analysis challenges in biological systems modeling, Industrial & Engineering Chemistry Research 48 (2009) 7168e7180.

34. S. Mishra, N. Deeds, G. Ruskauff, Global sensitivity analysis techniques for probabilistic ground water modeling, Ground Water 47 (2009) 730e747.

35. A. Saltelli, Making best use of model evaluations to compute sensitivity indices, Computer Physics Communications 145 (2002) 280e297

Comparative Scuffing Performance and Chemical Analysis of Metallic Surfaces for Air-Conditioning Compressors in the Presence of Environmentally Friendly Co$_2$ Refrigerant

Emerson Escobar Nunez, Nicholaos G. Demas,
Kyriaki Polychronopoulou,
and Andreas A. Polycarpou

Department of Mechanical Science and Engineering, 1206 West Green St., University of Illinois at Urbana-Champaign, Urbana, IL 61801, USA

ABSTRACT

Carbon dioxide (CO_2) has received significant interest as an alternative refrigerant for air-conditioning compressors due to its environmental benefits. These environmental benefits include zero ozone depletion potential and minimal global warming potential compared to commonly used hydrochlorofluorocarbon and hydrofluorocarbon refrigerants. This study presents results for three typical metallic tribopairs commonly found in air conditioning compressors, namely, Al390-T6, gray cast iron, and Mn–Si brass against 52,100 steel pins. The experiments were performed using a specialized tribometer capable of simulating compressor conditions, and in the presence of CO_2 and polyalkylene glycol lubricant. It was found that the scuffing resistance of gray cast iron and Mn–Si brass was similar and both materials performed better than Al390-T6. Through scanning electron microscopy and energy dispersive spectroscopy it was found that lead in Mn–Si brass melted during scuffing, and prevented sudden catastrophic failure of Mn–Si brass, unlike gray cast iron and Al390-T6 which failed abruptly. X-ray photoelectron spectroscopy conducted on the worn surfaces showed that chemically different species were present on the surfaces and their lubricious effect, originating from different metal oxides, could explain the scuffing behavior of the investigated alloys.

INTRODUCTION

Since the early 2000s the air-conditioning industry has been focusing on alternative refrigerants for the replacement of hydrochlorofluorocarbon (HCFC) and hydrofluorocarbon (HFC) refrigerants due to environmental regulations. The interest for long term solutions has been towards natural environmentally friendly refrigerants. Among different natural refrigerants such as water, air, isobutene (R600a), and ammonia, CO_2 (R744) is an attractive candidate [1]. CO_2 has been proven to be nontoxic and nonflammable with zero ozone depletion potential and negligible global warming potential compared to HCFC and HFC refrigerants [2]. However, its implementation in air-conditioning systems has been difficult because CO_2 systems have to be operated at high system pressures [3].

In air-conditioning compressors the solubility between refrigerant and lubricant plays an important role as the mixture circulates throughout the system and returns back into the compressor. Circulation has to be ensured to lubricate the tribopairs in the compressor and avoid pressure drops inside the system [4]. Materials such as Al390-T6, gray cast iron, and Mn–Si brass (UNS C67300) are commonly used for critical tribopairs in air-conditioning compressors, with Al390-T6 and Mn–Si brass found in automotive air-conditioning compressors and gray cast iron in industrial scroll compressors.

The wear and scuffing performance of silicon enriched aluminum alloys, such as Al390-T6, have been investigated before [5], [6], [7] and [8] with the main emphasis being on the tribological performance leading to scuffing. In the case of unlubricated experiments [5], it was shown that scuffing was caused by plastic flow and propagation of voids and cracks from the subsurface leading to the removal of a transformed top layer of material. Similar conclusions were reached for boundary/mixed lubricated experiments.

Several research groups have reported on the tribological performance of gray cast iron. Ref. [9] reported a positive effect of phosphorus and boron addition on the wear resistance of pearlitic gray cast iron. Good tribological behavior was also reported when gray cast iron was tested against gray cast iron in a CO_2 atmosphere as compared to R134a using PAG lubricant under boundary/mixed lubricated conditions [10]. An increase in the friction coefficient was also reported with an increase in the percentage of graphite phase in gray cast iron disks (tested against linings containing steel fibers) [11]. In this case it was proposed that the friction coefficient increased due to the interaction between sharp corners of the graphite flakes and the steel fibers present in the linings.

Mn–Si brass is a high tensile brass commonly used for applications where good resistance and low friction coefficient is required as for example in automotive synchronizers in gearboxes, where the tribological performance was reported to depend on hardness and relative fraction of its α and β phases [12]. It was reported that wear originated from the detachment of hard Mn_5Si_3 particles which resulted in material weakening. In a separate study it was reported that an increase in the α phase increased the wear resistance [13].

The objective of this study is to perform a comparative investigation of the tribological scuffing performance of Al390-T6, gray cast iron, and Mn–Si brass in CO_2 refrigerant. Compressor conditions were simulated using a specialized high pressure tribometer [14] under boundary/mixed lubrication conditions using PAG lubricant. The scuffed surfaces were analyzed using scanning electron microscopy (SEM), electron dispersive spectroscopy (EDS), and X-ray photoelectron spectroscopy (XPS) to identify the chemical nature of the formed tribolayers and their lubricity effect.

EXPERIMENTAL PROCEDURE

Controlled Tribological Experiments

Using a specialized high pressure tribometer [6], [14] and [15] in pin-on-disk configuration, a set of scuffing experiments were performed where the scuffing resistance was determined by progressively increasing the normal load up to the point of failure. The PAG lubricant (Idemitsu Kosan Co., Ltd., PZ 68ZL) used in this study was specifically manufactured for use with CO_2.

Experiments were performed on Al390-T6, gray cast iron, and Mn–Si brass disks tested against swash plate compressor 52,100 steel pins. The tests were performed at room temperature (22 °C) to minimize viscosity changes with temperature. Before each experiment a small amount of 40 mg of PAG lubricant (approximately 2 drops) was applied onto the surface of the pins. The normal load was increased in steps of 67 N every 15 s up to the point of scuffing. A rotational speed of 1000 rpm, corresponding to a linear speed of 2.4 m/s was used and the CO_2 chamber pressure was kept constant at 0.17 MPa. During the experiments the scuffing point was characterized by a sudden increase in the friction coefficient manifested through the formation of cold welds between the pin and the disk.

The chamber pressure used during the experiments is lower than typical pressures experienced by CO_2 compressors, which are around 3.0 MPa and 12.0 MPa for the compressor low and high pressure sides, respectively. The reason for using lower chamber pressures, other than the fact that is convenient, is justified since during boundary/mixed lubrication experiments using CO_2 refrigerant and PAG lubricant, the effect of CO_2 pressure on viscosity is relatively small. This can be explained by the partial solubility between PAG lubricant and CO_2. Such measurements reported in Refs. [17] and [18] show that the amount of CO_2 that can be dissolved in PAG lubricant is limited and the viscosity of PAG lubricant does not decrease significantly as a function of the CO_2 pressure (keeping the temperature approximately constant). Furthermore, the selection of low chamber pressure can be justified based on earlier tribological studies [3] at high CO_2 working pressures (boundary/mixed lubrication conditions in the presence of PAG) where no significant differences were found in terms of friction coefficient and wear after testing at CO_2 chamber pressures of 1.4, 4.1, and 6.9 MPa.

In past experiments under similar operating conditions [10], measurements of near contact temperature (subsurface temperature measured 2 mm below the sliding interface) showed that the temperature remained steady after the running-in period and only showed a sudden increase at the onset of scuffing. This suggests that after the running-in period, the lubricant film and CO_2 lubricity become stable thus significantly reducing asperity interactions and thus high flash temperatures, which could affect the lubricant viscosity.

Before each experiment the samples were immersed in a pool of acetone and ultrasonically cleaned, then rinsed with alcohol and dried using warm air. To ensure repeatability, each experiment was performed at least twice. In the contact geometry used for these experiments, the pin was the lower stationary part, and the disk was the upper rotating part. Photographs of typical tested samples are shown in Fig. 1. Note that the 52,100 steel pins are curved or "crowned" with a radius of curvature of 0.3 m and also have a dimpled geometry that helps retain lubricant during sliding [16].

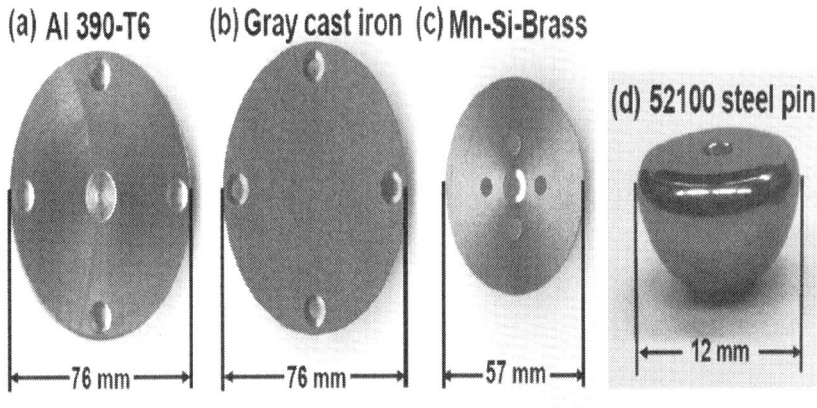

Figure 1: Samples used for the scuffing experiments (a)–(c) disk materials and (d) crowned steel pin.

To obtain similar surface roughness, all disks were machined and polished using the same technique, namely lapping. The root-mean square roughness of the samples was determined using 10 mm and 1 mm long profilometric scans for the disks and pins, respectively. After testing, cross-sections of Al390-T6, gray cast iron, and Mn–Si brass were prepared by cutting the samples exposing the cross-section of the wear tracks, then polishing using different grit emery papers ranging in ANSI standard grit size from 320 to 4000. Final mirror polishing was performed using an emery cloth. The samples were used to take SEM cross-section images.

EDS

EDS experiments were conducted on a high resolution, field emission JEOL 7000F SEM. The electron source is a Schottky field emission gun with probe current of 1 pA to 200 nA and accelerating voltage from 0.5 kV to 30 kV. EDS studies were performed inside the wear tracks to analyze chemical composition.

XPS

XPS experiments were conducted on a KRATOS spectrometer equipped with a hemispherical electron analyzer and a non-monochromatic Al K X-ray source (1253.6 eV). All reported photoelectron binding energies are referenced to the C 1s feature of adventitious carbon at 284.6 eV to take into account charging effects. XPS studies were performed both inside and outside the wear tracks after the scuffing experiments. A certain region of the spectrum was scanned a number of times in order to obtain a good signal-to-noise ratio using a pass energy of 40 eV. The measurements were performed in three different regions inside the wear track in each individual sample for repeatability.

RESULTS AND DISCUSSION

Microstructural Study of the Untested Samples

Al390-T6

The chemical composition of the aluminum alloy used in this study is shown in Table 1[16], indicating that there is a high percentage of silicon (16–18%), which provides good wear resistance. As seen in the SEM microphotograph and EDS mapping depicted in Fig. 2a and b, respectively, silicon particles in Al390-T6 have a non-uniform morphology: There are primary silicon particles (large dark particles) and eutectic silicon particles (small dark particles) embedded in the Al matrix (Fig. 2c). Primary silicon particles are responsible for the improvement of the strength of the material and wear resistance, and as measured in[16] and [19], the hardness of pure silicon particles is 10–12 GPa. The eutectic silicon particles have a more uniform distribution in the matrix than the primary silicon particles, as seen in Fig. 2a and b. Additional elements like copper (Cu) and magnesium (Mg) forming different phases as seen in Fig. 2d and e provide additional strength to the alloy. Fe and Mn are not observed because their percentages are too low to be identified through EDS.

Table 1: Chemical composition of Al390-T6, gray cast iron, and Mn–Si brass (wt.%).

Element	Specified chemical percentage by weight (wt%)		
	Al390-T6	Gray cast iron	Mn–Si brass
Al	76.00	–	0.01
C	–	3.20–3.70	–
Cu	3.00–4.00	–	60.40
Fe	1.00	Balanced	0.23
Mg	0.40–1.00	–	–
Mn	0.50	0.70–0.80	2.43
Pb	–	–	1.24
Si	16.00–18.50	2.20–2.55	0.86
Zn	1.00	–	35.58

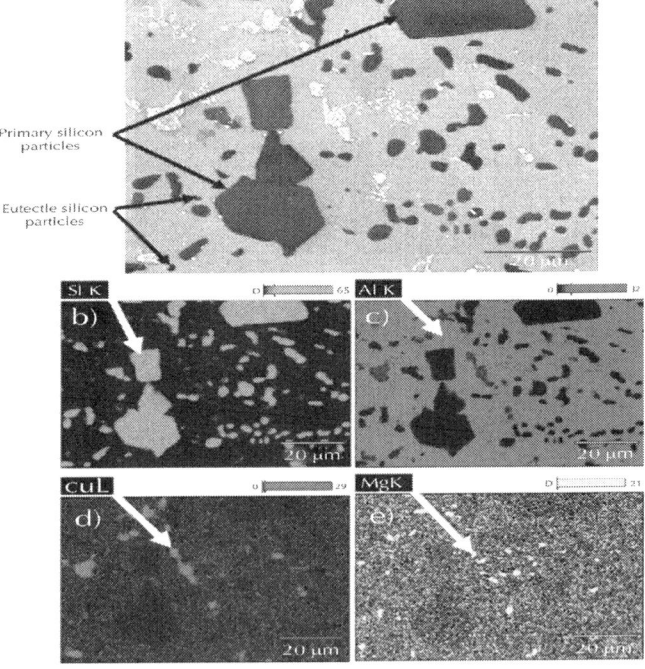

Figure 2: (a) Surface cross-section SEM micrograph of untested Al390-T6 hypereutectic alloy showing primary and eutectic silicon particles, and (b)–(e) EDS mapping showing Si, Al, Cu, and Mg respectively.

Gray Cast Iron

Gray cast irons are formed by the addition of silicon, which promotes the formation of graphite flakes when its concentration is greater than 1%. Table 1 also shows the chemical composition of gray cast iron [6], while Fig. 3a and b shows the microstructure of the pearlitic gray cast iron used in this work. It can be seen that graphite flakes have sharp edges and are surrounded by pearlitic structure where cementite and ferrite appear as light lamellar structure and dark structure, respectively.

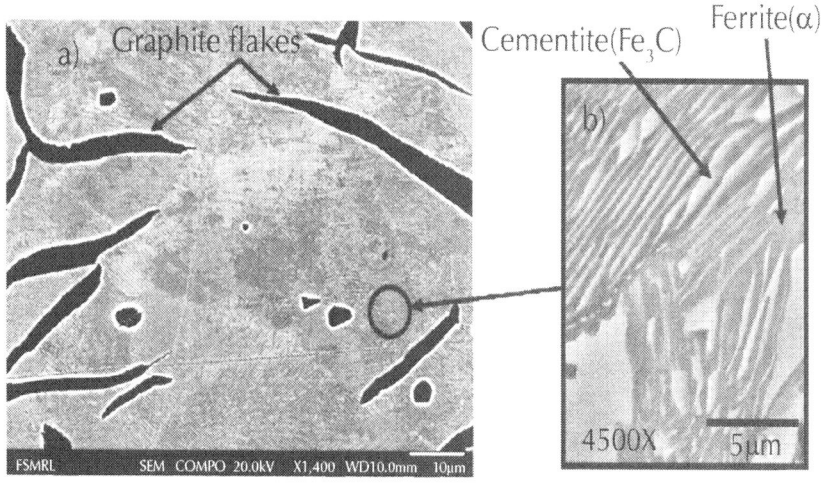

Figure 3: Surface cross-section SEM micrograph of untested pearlitic gray cast iron: (a) graphite flakes surrounded by pearlitic matrix and (b) cementite (Fe_3C) light lamellar structure and ferrite (α) dark structure.

Mn–Si brass

The microstructure of Mn–Si brass consists of a copper rich α soft phase matrix and a manganese silicide phase (Mn_5Si_3) [20] (its composition is also listed in Table 1). The low percentage of Al (less than 0.02%) causes the matrix to be rich in α phase. Silicon increases the wear resistance of Mn–Si brass by forming hard Mn_5Si_3 and improves its performance during plastic deformation [21]. Primary (larger) and smaller Mn_5Si_3 particles can be seen in Fig. 4 along with lead (Pb).

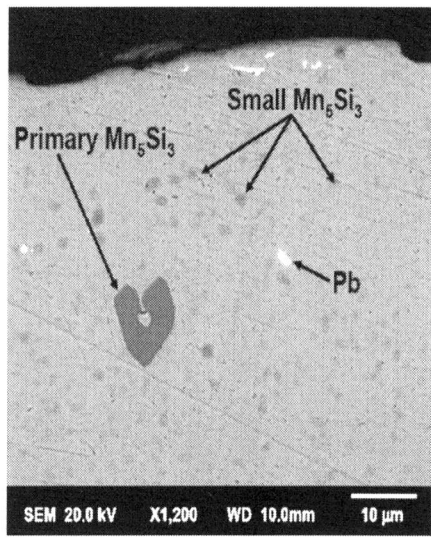

Figure 4: Surface cross-section SEM micrograph of untested Mn–Si brass showing primary and small Mn_5Si_3 particles and lead (Pb).

Controlled Tribological Scuffing Experiments

Typical scuffing experimental results are shown in Fig. 5, Fig. 6 and Fig. 7, where the in situ normal load and friction coefficient are shown for the three different material pairs. Before the experiments, the surface roughness were measured and found to be 0.7 μm and 20 nm for the disks and pins, respectively. As seen in Fig. 5, the scuffing resistance (maximum normal load to cause scuffing) of Al390-T6 was approximately 500 N. For gray cast iron and Mn–Si brass, the scuffing resistance was higher at 850 N as shown in Fig. 6 and Fig. 7 respectively. The tribological behavior of Mn–Si brass near scuffing was different compared to Al390-T6 and gray cast iron, as it failed gradually. This phenomenon can be attributed to the presence of Pb in Mn–Si brass, which has a low melting point. It has been reported that friction coefficient and temperature increase abruptly during scuffing [22] and the increase in temperature can cause melting of Pb at the onset of scuffing allowing the Pb to act as a lubricant. It is believed that this is the reason why under the presence of a sudden transient load Mn–Si brass can resist catastrophic failure as can be seen in Fig. 7.

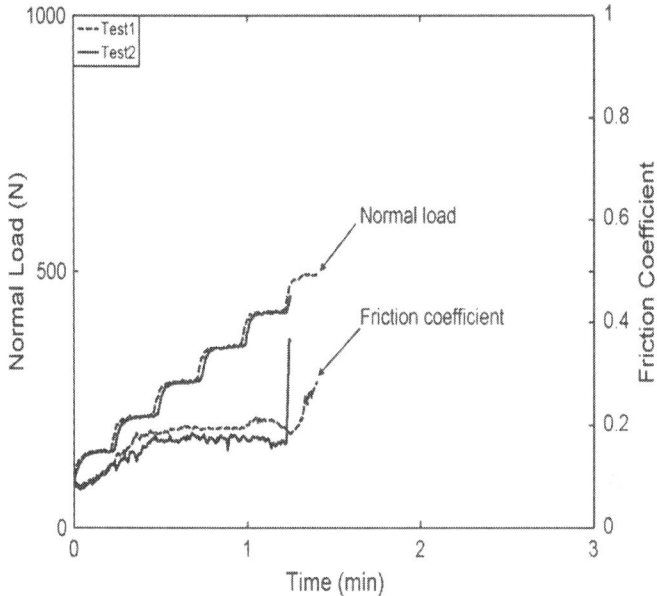

Figure 5: Typical scuffing experiments for Al390-T6.

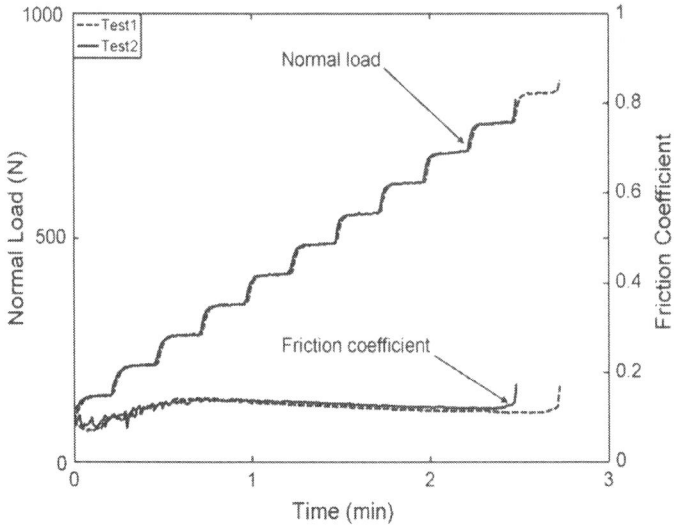

Figure 6: Typical scuffing experiments for gray cast iron.

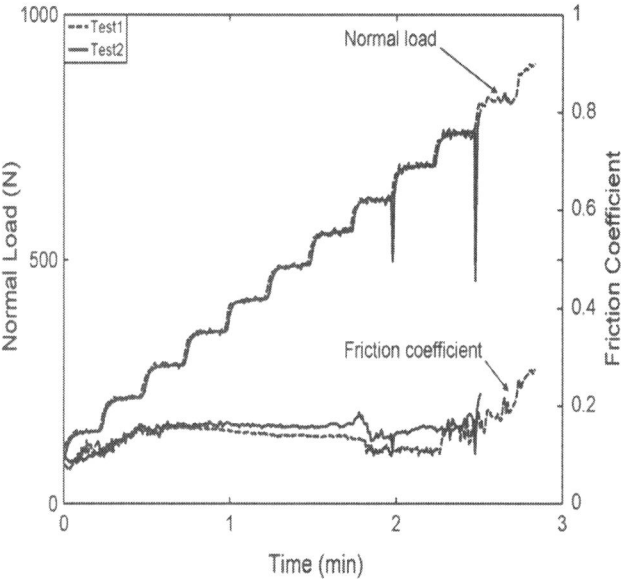

Figure 7: Typical scuffing experiments for Mn–Si brass.

The friction coefficient values were similar in all cases with Al390-T6 being somewhat higher (~0.18) compared to gray cast iron and Mn–Si brass (~0.14). Despite the fact that these experiments are short in duration, they provide information about the maximum load sustained before failure (scuffing). From these experiments it is evident that gray cast iron and Mn–Si brass perform better than Al390-T6 in terms of both scuffing resistance and lower friction coefficient values. Also, it should be noted that in these experiments the friction coefficient was unsteady at the beginning of the tests due to run-in, which was similar for the three materials as the same surface preparation was used to prepare the samples.

SEM Studies of Scuffed Disks

SEM images of the wear tracks are shown in Fig. 8. Referring to Fig. 8a, Al390-T6 shows large delaminated regions in the scuffed surface, while Fig. 8b shows typical galling behavior. Failure of the lubricant film causes the destruction of the protective tribolayer [5], thus causing scuffing. With scuffing, fracture of subsurface primary silicon particles is evident, as shown in Fig. 9, thus exposing hard and abrasive silicon particles at the interface. Eventually, severe adhesion takes place leading to the plastic shearing of the bulk of the material (seizure). This explains the large delamination bands present in the Al390-T6 sample after scuffing.

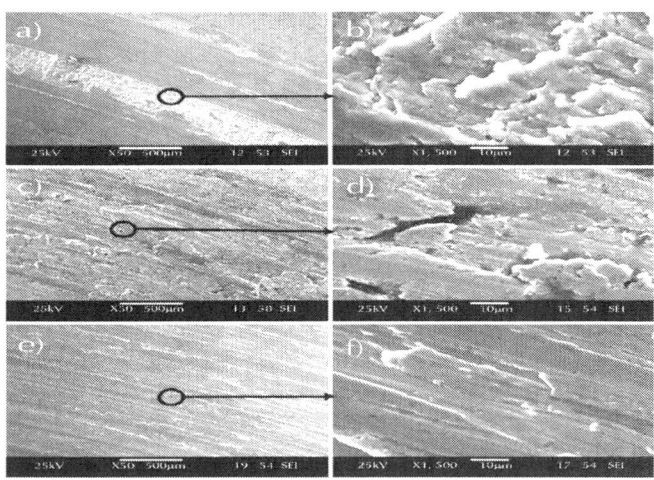

Figure 8: SEM surface images inside the wear track of (a) and (b) Al390-T6, (c) and (d) gray cast iron and (e) and (f) Mn–Si brass after scuffing.

Figure 9: Surface cross-section SEM image of Al390-T6 showing cracks on a primary silicon particle below the surface during scuffing.

In the case of gray cast iron shown in Fig. 8c and d less adhesive wear compared to Al390-T6 was observed. Due to the presence of graphite flakes, a sudden failure can be expected during scuffing because the tips of the graphite flakes are sharp thus acting as stress concentration sites, as also shown inFig. 10a and b. Mn–Si brass shown in Fig. 8e and f shows minor adhesive wear with plastic deformation during scuffing. Large plastic deformation during scuffing in the case of Mn–Si brass can be explained by the detachment and fracture of hard Mn_5Si_3 particles during sliding, as shown in Fig. 11. Since the size and percent by volume of Mn_5Si_3 particles is small, it is believed that after fracture, depletion of these particles leads to a decrease in strength and fatigue resistance leaving a softer matrix that can more readily flow plastically, as seen in Fig. 8f.

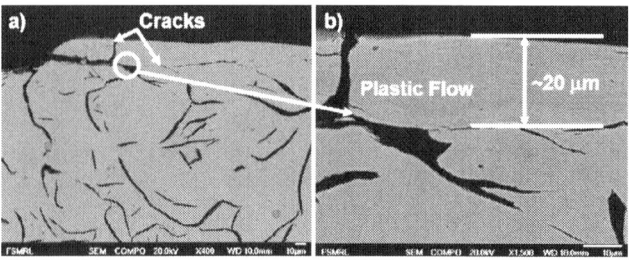

Figure 10: Surface cross-section SEM image of gray cast iron showing plastic flow and propagation of cracks originated from graphite flakes below the surface during scuffing.

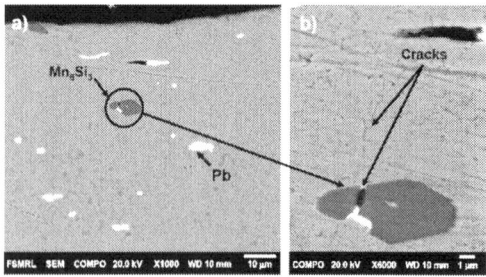

Figure 11: Surface cross-section SEM image of Mn–Si brass showing a fractured Mn_5Si_3 particle and cracks propagating from this particle during scuffing.

EDS Studies

SEM cross section analysis of scuffed Mn–Si brass is depicted in Fig. 12 where the spectra and the chemical elements at each point inside the wear track are presented in Fig. 13. Fracture of Mn_5Si_3 particles and propagation of cracks resulted in surface failure. Spectral analysis reveals the presence of lead inside the broken Mn_5Si_3 particles (spectrums 2 and 4 in Fig. 13). This finding supports the observation made previously that Pb melts during scuffing of Mn–Si brass leading to a gradual failure. The local melting of Pb and thus its lubricity effect is due to an increase in temperature at the onset of scuffing.

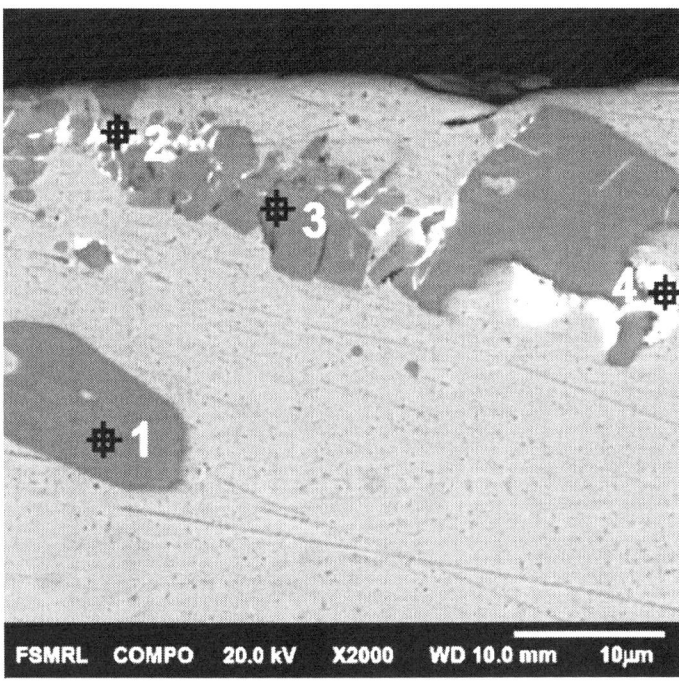

Figure 12: Surface cross-section SEM micrograph inside the wear track of Mn–Si brass (numbers refer to EDS spectra shown in Fig. 13).

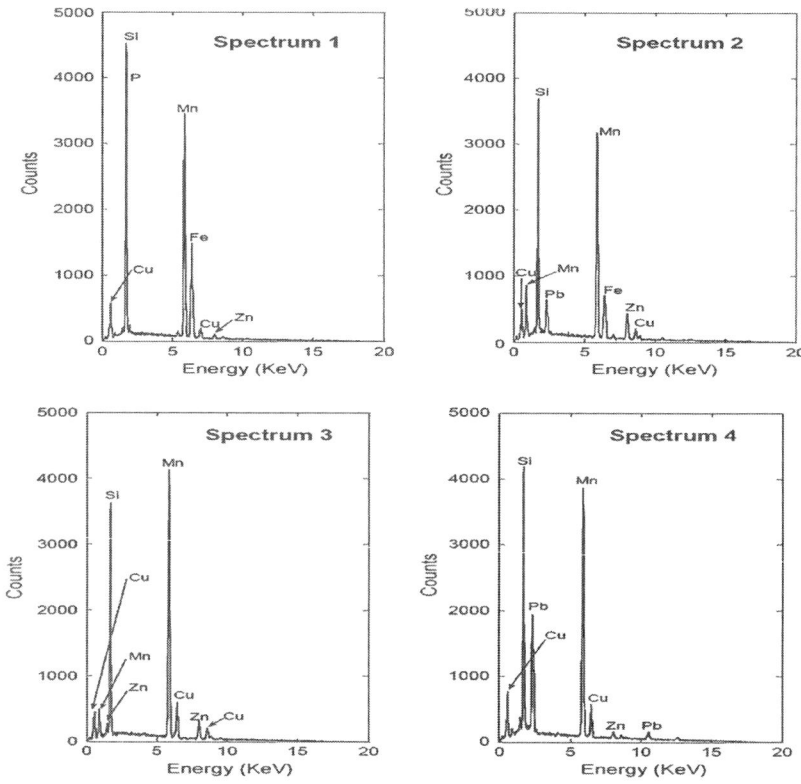

Figure 13: EDS spectra displaying the chemical composition from cross-section SEM of Mn–Si brass (refer to Fig. 12).

XPS Studies

The XPS C 1s and O 1s core level spectra for the worn surfaces of Al390-T6, gray cast iron, and Mn–Si brass are presented in Fig. 14a and b, respectively. C 1s XPS peak in the case of Al390-T6 consists of a symmetric peak, whereas in the case of Mn–Si brass and gray cast iron more carbon components can be seen. The peak at 284.6 eV corresponds to the C contamination, which is always present [23] and [24]. The C 1s peak at 287.1 corresponds to –(C–O–R or C=O) entities, the origin of which could be from the fragmentation of the PAG lubricant used in the experiments [23] and [24]. Due to polar oxygen species, this layer exhibits increased adhesion to the metallic surface.

It is suggested that the good chemical affinity (improved adhesion) of this transfer layer with Mn–Si brass and gray cast iron might be the reason of their better performance, unlike the case for Al390-T6 alloy, in which case this transfer layer seems not to be XPS detectable. In the case of Al390-T6, the XPS investigation inside the wear track showed the presence of fluorine (which is probably a lubricant additive) something that at a first sight seems to be opposite to the previous findings, though this could imply that indeed a transfer layer exists in the case of Al390-T6 alloy, but the chemical composition of this layer is different due to different interaction of the lubricant/additives with the metal aluminum surface. This different chemical identity of the transfer layer could lead to different wear mechanisms. No fluorine was detected in the cases of gray cast iron and Mn–Si brass materials.

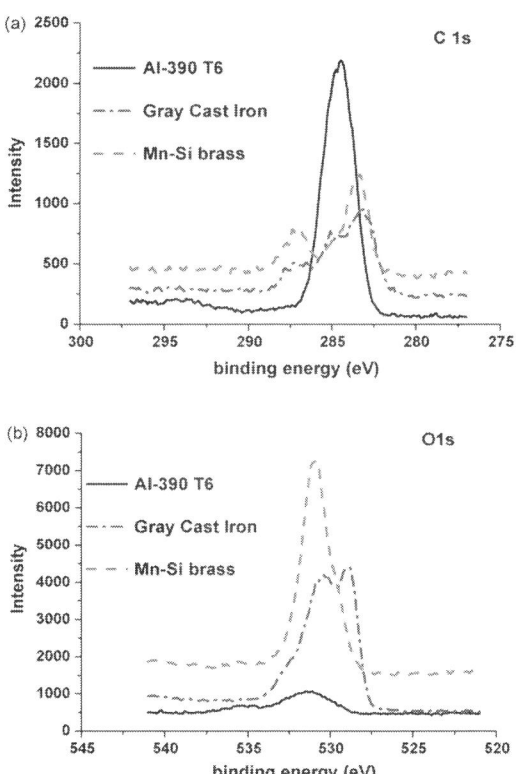

Figure 14: (a) C 1s and (b) O 1s XPS core level spectra obtained inside the wear track of Al390-T6, Gray cast iron, and Mn–Si brass.

Fig. 14b presents the O 1s core level spectra. In the case of Al390-T6, a small broad peak centered at 531.1 eV is seen, which corresponds to O$^-$ adsorbed on the surface (most probably coming from CO_2). In the case of gray cast iron a peak at 528.9 eV corresponding to metal oxide formed (most probable iron oxide due to the iron abundance in the alloy), a peak at 530.5 eV which corresponds to O$^-$ adsorbed on the surface and a shoulder at 532.6 eV which corresponds to C–O–C (ether-type oxygen species) coming from the lubricant used [23] and [24]. In the case of Mn–Si brass the O 1s peak appears at 531.1 eV and corresponds to adsorbed oxygen, whereas the contribution coming from a metal oxide at 529.3 eV appears as a shoulder in this case. The oxide in this case could be copper oxide due to the abundance of copper in this alloy. In the case of gray cast iron, the peak at 531.1 eV present lower intensity and also another component at 528.9 eV appeared, which corresponds to metal oxide. It is known that graphite has a layered structure with carbon atoms bond to each other in the layer by strong covalent bonds. Under applied tribological conditions breakdown of graphite structure can occur leading to species formation which contribute to the peak at 531.1 eV. The fragmentation of graphite leads to the formation of exposed covalent bonds and wear debris, entities having strong adsorption tendency (chemical affinity) to the oxygen (present in air) and H_2O-originating groups (such as OH) (present in a humid environment) leading thus to species corresponding to the peak at 531.1 eV as previously mentioned [25].

In the above cases of oxides formed inside the wear track of cast iron and Mn–Si brass alloys the possibility of binary oxides formation cannot be excluded. The different metal cations in the two alloys could lead to a totally different combination of oxides which seems to be beneficial in the case of Mn–Si brass in addition to the previously discussed Pb melting positive effect. This could be attributed to the fact that Mn–Si brass mainly contains Cu (major element), the oxide of which has been reported to exhibit lubricious behavior, in comparison with iron oxide (oxide formed in the case of cast iron) [26]. Also in the case of Mn–Si brass, Cu_2O and CuO have been found according to our XPS studies. Based on the crystal chemistry approach introduced by Erdemir et al. [26] and [27], in the case of binary oxides as the difference in ionic potential between the two metal cation centers increases, the formed oxides (under tribo-testing conditions) exhibit an increased tendency of forming shearable species [27]. This has the consequence of lowering the hardness and shear strength at high temperatures. According to this approach, a better screening of cations by the anions, which potentially leads to less interaction between the former and in turn leads to lower friction, can also explain the better performance of gray cast iron and Mn–Si brass compared to Al390-T6, based dominantly on the nature of different oxides formed in the contact zone under tribological conditions applied.

CONCLUSIONS

A comparative tribological performance of three different materials used in air-conditioning compressors and in the presence of the environmentally friendly refrigerant CO_2 was carried out. Results showed that the scuffing performance of gray cast iron and Mn–Si brass was similar and both materials performed better than Al390-T6. Unlike cast iron and Al390-T6 that exhibited abrupt scuffing failures, Mn–Si brass exhibited gradual failure, which could be advantageous in actual applications. This is an interesting finding taking into account that gray cast iron has a higher strength compared to Mn–Si brass. Based on EDS surface analysis, retardation of scuffing in the case of Mn–Si was attributed partly to the presence of lead in the matrix which melted during the onset of scuffing providing additional lubricity. Through SEM it was found that in gray cast iron cracks nucleated and propagated in the sharp corners of graphite flakes at the onset of scuffing. Also, using SEM it was shown that subsurface failure in Mn–Si brass was from fragmentation of Mn_5Si_3 particles, which was caused by stress raisers in the sharp corners of these particles leading to material failure. In the case of Al390-T6 subsurface failure was initiated in the sharp corners of primary Si particles leading to its depletion to the surface and thus weakening of this alloy. XPS showed that there is a correlation between different species coming from the fragmentation of the PAG lubricant and the scuffing behavior of the different materials tested. Also, different metal oxides formed on the surfaces could have a beneficial lubricious effect contributing to an increased scuffing resistance, as seen with Mn–Si brass and gray cast iron.

ACKNOWLEDGEMENTS

This research work was supported by the 29 member companies of the Air-Conditioning and Refrigeration Center, an Industry-University Cooperative Research Center at the University of Illinois at Urbana-Champaign. Chemical analyses were performed at the Center for Microanalysis of Materials, University of Illinois at Urbana-Champaign, which is partially supported by the U.S. Department of Energy under grantDEFG02-91-ER45439.

REFERENCES

1. M.H. Kim, J. Pettersen, C.W. Bullard, Fundamental process and system design issues in CO2 vapor compression systems, Prog. Energy Combust. Sci. 30 (2004) 119–174.

2. G. Lorentzen, The use of natural refrigerants: a complete solution to the CFC/HCFC predicament, Int. J. Refrig. 18 (1995) 190–197.

3. N.G. Demas, A.A. Polycarpou, Ultra high pressure tribometer for testing CO2 refrigerant at chamber pressures up to 2000 psi to simulate compressor conditions, Tribol. Trans. 49 (2006) 1–6.

4. A. Yokoseki, Solubility correlation and phase behaviors of carbon dioxide and lubricant oil mixtures, Appl. Energy 84 (2) (2006) 159–175.

5. T. Sheiretov, H. Yoon, C. Cusano, Scuffing under dry sliding conditions—part I.Experimental studies, Tribol. Trans. 41 (1998) 435–446.

6. A.Y. Suh, J.J. Patel, A.A. Polycarpou, T.F. Conry, Scuffing of cast iron and Al390-T6 materials used in compressor applications, Wear 260 (2006) 735–744.

7. H. Yoon, T. Sheiretov, C. Cusano, Scuffing behavior of 390 Aluminum against steel under starved lubrication conditions, Wear 237 (2000) 163–175.

8. D.K. Dwivedi, Wear behavior of cast hypereutectic aluminum silicon alloys, Mater. Des. 27 (2006) 610–616.

9. J. Keller, V. Fridrici, P. Kapsa, S. Vidaller, J.F. Huard, Influence of chemical composition and microstructure of gray cast iron

on wear of heavy duty diesel engines cylinder liners, Wear 263 (2007) 1158–1164.

10. N.G. Demas, A.A. Polycarpou, Tribological studies on scuffing due to the influence of carbon dioxide used as a refrigerant in compressors, Tribol. Trans. 48 (2005) 336–342.

11. M.H. Cho, S.J. Kim, R.H. Basch, J.W. Fash, H. Jang, Tribological study of gray cast iron with automotive brake lining, Tribol. Int. 36 (2003) 537–545.

12. A. Waheed, N. Ridley, Microstructure and wear of some high-tensile brasses, J. Mater. Process. Technol. 114 (1994) 201–211.

13. H. Mindivan, H. C¸ imenoglu, E.S. Kayali, Microstructure and wear of brass syn- ˜ chronizer rings, Wear 254 (2003) 532–537.

14. T. Sheiretov, W.V. Glabbeek, C. Cusano, Evaluation of the tribological properties of polyimide and poly(amine-imine) polymers in a refrigerant environment, Tribol. Trans. 38 (1995) 914–922.

15. A.Y. Suh, A.A. Polycarpou, T.F. Conry, Detailed surface roughness characterization of engineering surfaces undergoing tribological testing leading to scuffing, Wear 255 (2003) 556–568.

16. S.R. Pergande, A.A. Polycarpou, T.F. Conry, Nanomechanical properties of aluminum 390-T6 rough surfaces undergoing tribological testing, J. Tribol. 126 (2004) 573–582.

17. C. Seeton, J. Fahl, D. Henderson, Solubility, viscosity, boundary lubrication and miscibility of CO2 and synthetic lubricants, in: Proceedings of the 4th IIR-Gustav Lorentzen Conference on Natural Working Fluids, West Lafayette, US, 2000, pp. 446–454.

18. A. Hauk, E. Weidner, Thermodynamic and fluid-dynamic properties of carbon dioxide with different lubricants in cooling circuits for automobile application,Ind. Eng. Chem. Res. 39 (2000) 4646–4651.

19. B.S. Shabel, A.G. Douglas,W.G. Truckner, Friction and wear of aluminum–silicon alloys, ASM Handbook 18 (1992) 785–794.

20. Y.S. Sun, G.W. Lorimer, N. Ridley, Microstructure of high-tensile strength brasses containing silicon and manganese, Metall. Mater. Trans. A 20A (1989) 1199–1206.

21. N.B. Pugacheva, Structure of commercial + brasses, Metal Sci. Heat Treat.49 (1–2) (2007) 67–74.

22. M.P. Cavatorta, C. Cusano, Running-in of aluminum/steel contacts under starved lubrication. Part II. Effects on scuffing, Wear 242 (2000) 133–139.

23. J. Chastain Jr., R.C. King, Handbook of X-ray Photoelectron Spectroscopy, Physical Electronics, Inc., MN, 1995, ISBN 0-9648124-1-X.

24. http://srdata.nist.gov/xps/ (accessed 12.08.09).

25. http://www.tribology-abc.com/abc/solidlub.htm   (accessed 17.08.09).

26. A. Erdemir, A crystal-chemical approach to lubrication by solid oxides, Tribol. Lett. 8 (2000) 97–102.

27. A. Erdemir, S. Li, Y. Jin, Relation of certain quantum chemical parameters to lubrication behavior of solid oxides, Int. J. Mol. Sci. 6 (2005) 203–218.

7

Fracture Analysis on the 4th Compressor Disc of Some Engine

x

X.L. Liu, W.F. Zhang, T. Jiang, and C.H. Tao

Failure Analysis Center, AVIC, Beijing Institute of Aeronautical Materials, Beijing 100095, China

ABSTRACT

Some plane was caused fire and exploded because of the fracture of the 4th disc of the engine. In this paper, in order to give the fracture mechanism and causes of the 4th disc some observation and analysis were made, such as the visual inspection of some debris of the 4th disc, the macroscopic and microscopic observation of the fracture surface on the flange and web of the 4th disc. The corrosion pits on the web surface and at the edge of the fracture surface were analyzed by energy spectrum. The beach marks on the fracture surface were measured. The material quality of the web was also analyzed. The vibration peculiarity of the engine was tested and calculated. The fracture mechanism of the 4th compressor disc is high cycle fatigue. The microstructure, hardness, chemical composition and the web thickness of the fracture site all meet with technical requirements. The fatigue failure has no connection with the material quality. The corrosion pits at the R3 arc transition between the web and the flange induced the accident. The cracking and propagation of the 4th compressor disc are mainly related to the vibration stresses.

INTRODUCTION

Ground-run test for some fighter was performed after a hydraulic pressure system fault was eliminated. Rested at low velocity state for 4 ~ 5 min, the engine was accelerated to $n_1 = 93\%$. Two minutes later, the plane caused fire and exploded. On the basis of local accident investigation, the experts affirmed that the fracture of 4th compressor disc resulted in the uncontained accident. That was to say, the 4th disc flange with blades firstly fractured from the disc, then penetrated through the casket and hit on the left wing gasoline tank, and finally built a fire and led to explosion.

Made by 1Cr11Ni2W2MoV alloy, the disc was quenched (1010 °C) and tempered (560 °C) after certain being founded and machined.

In this paper, the fracture mechanism and failure causes of the 4th disc are discussed and determined.

EXPERIMENTAL PROCEDURE

The visual inspection of some debris of the 4th disc was made, and there were fatigue characteristics on the fracture surface of R3 transitional arc between the flange and the web. The fatigue area and the direction of crack propagation were approximation confirmed. The fatigue fracture surface of the flange and the web was cut so as to be observed by scanning electron microscope (SEM). The origin site was made certain and the beach marks were measured by SEM. The corrosion matter on the origin site and on the web surface was analyzed by energy spectrum, and the corrosion medium was known. Metallurgical samples were removed from the web in parallel and vertical directions, respectively. The metallurgical examination, hardness test and chemical composition analysis of the material of web were made. The vibration peculiarity was improved using laboratory test and computer simulation.

EXPERIMENTAL RESULTS

Visual Inspection

Fig. 1 is part sketch of the 4th compressor disc, the crack is at R3 site, shown used arrowhead in Fig. 1. Fig. 2 shows the part debris with fatigue region of the fractured 4th disc flange. The flange was severely distorted and most of the blades fell off due to wear and distortion. There are abundant black corrosion spots and distinct machining marks on the inner surface of the R3 arc. The machining marks indicate that this position was not polished. The compressor disc web appeared red–black[1] which implied it has suffered fire. Furthermore, for the front side surface of the web, the color was light and there was no obvious machining marks. While for the rear side surface of the web, the color was dark and there were several apparent circumferential machining marks near the fracture surface, which propagated along one of the marks.

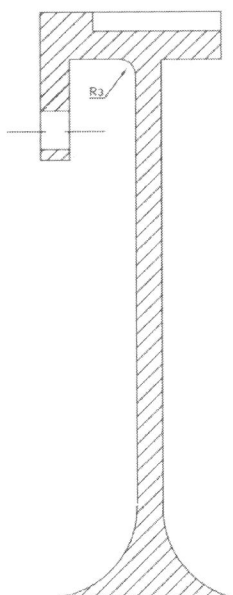

Figure 1: Part sketch of the 4th compressor disc.

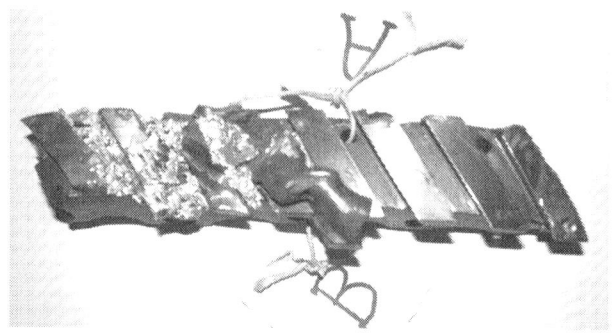

Figure 2: The appearance of the tenon groove of flange.

Surrounded by 45° inclined rapid fracture areas, a flat and arc area with a length of 62.5 mm on the fracture surface suggests it is a fatigue cracking, as shown in Fig. 3. The crack initiated at the R3 arc of the front side surface, then propagated along circumferential direction as well as to the rear side surface. Appeared yellow and black, the fracture surface nearly spanned the web thickness except a 45° inclined rapid fracture area with a maximum width of 0.2 mm.

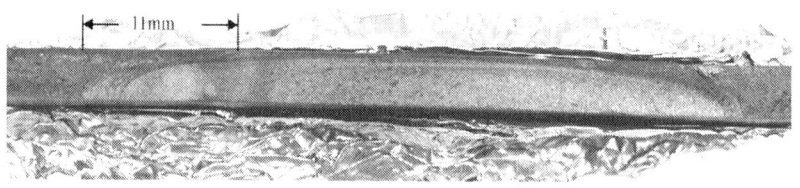

Figure 3: Fatigue fracture appearance at the edge of the web.

The cleaned fatigue fracture surface from the web gave evidence of fatigue features. It can be seen that the arc fatigue area is not centrosymmetric, while there are full centrosymmetric fatigue beach marks on the right area of the fracture surface. A judgement was made that the main origin is situated at the convergent center of the full beach marks. The distance is 19.5 mm from the main origin to the right rapid fracture area.

Another little fatigue area opposite the main fatigue area can be seen in Fig. 3. It originated from the rear side surface, extended to a length of 11 mm and a width of 1 mm.

Microscopic Observation and Energy Spectrum Analysis

The fatigue fracture surface of the wheel flange was cut to be observed by SEM. The origin can not be observed due to the damage and distortion at the crack edge, but can be inferred by the fatigue striations that it located at the front side and propagated to the back side. There are apparent beach marks and fine dense fatigue striations in the propagation anaphase area, as seen in Fig. 4. Although most of the fracture surface is badly scraped, some areas without scrape and corrosion signs reveal the crack propagated under mechanical stresses.

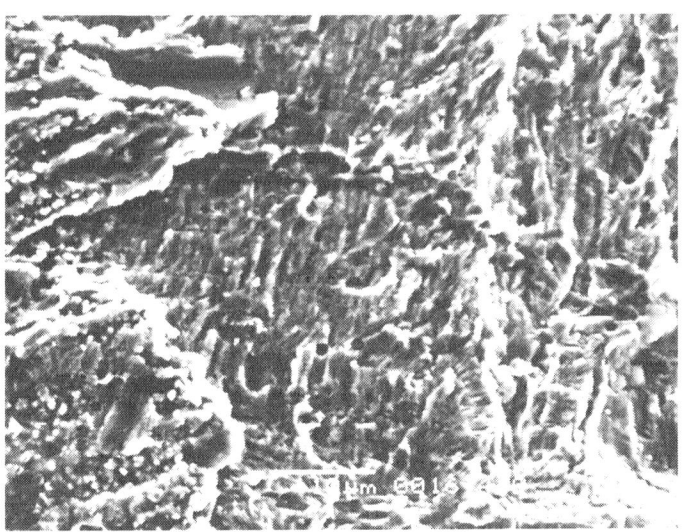

Figure 4: The beach marks and fatigue striation on fracture surface of the flange.

SEM observation for the 4th disc web without being cleaned shows that there are many pits with different size and shape, as shown in Fig. 5. At the edge of fatigue fracture surface, there are also some small quantity of corrosion pits. Energy spectrum analysis demonstrates that these pits are corrosion pits resulted from chlorine ion.

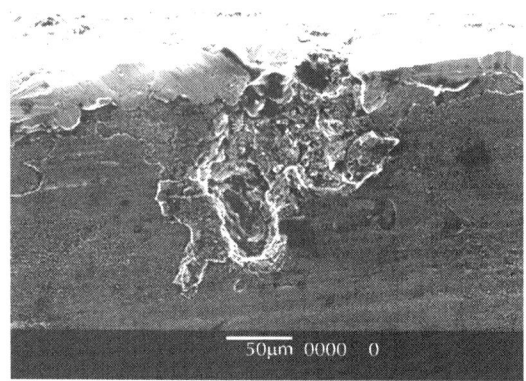

Figure 5: The corrosion pits at the edge of the flange surface.

A dense layer, covering on the fracture surface of web, is oxide and extraneous contamination, as proved by energy spectrum analysis. There is a pit at the edge of fatigue origin. Detailed observation indicates the fracture surface features are convergent to the point (Fig. 6), therefore, it is the point of crack initiation. The pit is approximately isosceles triangle with irregular edge shape, and there are grains in the pit. Energy spectrum analysis shows the grains are corrosion products and the pit is a corrosion pit, with 185 μm in length and 80.4 μm in depth.

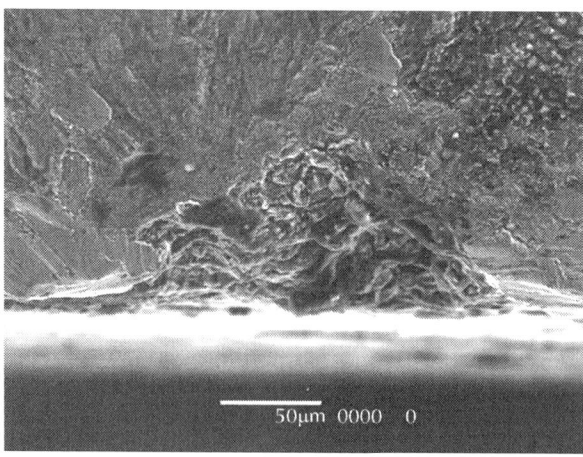

Figure 6: The fracture surface features are convergent to the point.

Table 1 shows the energy spectrum analysis results of different sites on the web.

Table 1: Foreign elements at various positions of the web before being cleaned (wt%)

Elements	O	Cl	K	Ca	Na	Mg	Al	
Pit on web surface	23.41	1.77	0.66	0.26	0.67	0.17	1.35	The rest is basic element
Pit at the edge of the fracture surface	9.37	0.17	0.14	0.20	0.27	0.19	0.13	
Covered layer on the fracture surface	14.38	0.20	0.13	–	–	–	1.18	
Pit at the fatigue origin	6.94	0.22	0.81	0.37	0.38	–	0.40	

The fracture surface of the web was observed after the surface had been rinsed by 50% positive H_3PO_4 solution and the oxidation film was removed. It is more obvious that the crack initiated at the corrosion pit because many fine radiative lines near the origin are convergent to the corrosion pit. The fatigue origin site is also flat. The little fatigue area opposite to the left side of the main fatigue region is a subordinate fatigue region, which originated from the rear side surface, extended to a length of 11 mm and a width of 1 mm, as can be seen in Fig. 3 and Fig. 7. On the subordinate fracture surface obvious beach marks can be seen, as shown in Fig. 7. It should be noted that the subordinate cracking originated from a line corresponding to machining marks, this provides the subordinate cracking is related to machining marks, as shown in Fig. 8.

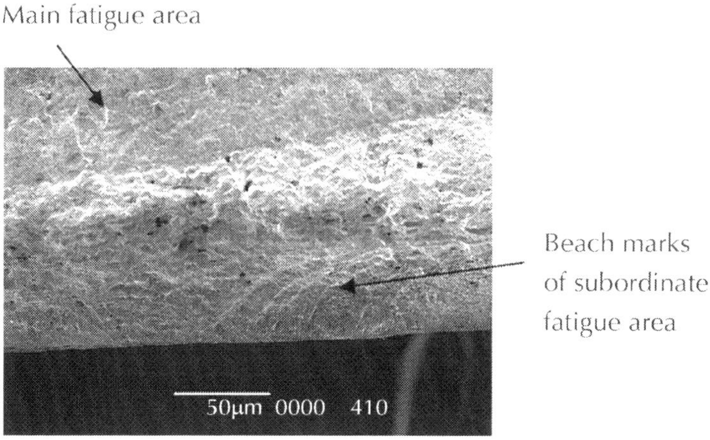

Main fatigue area

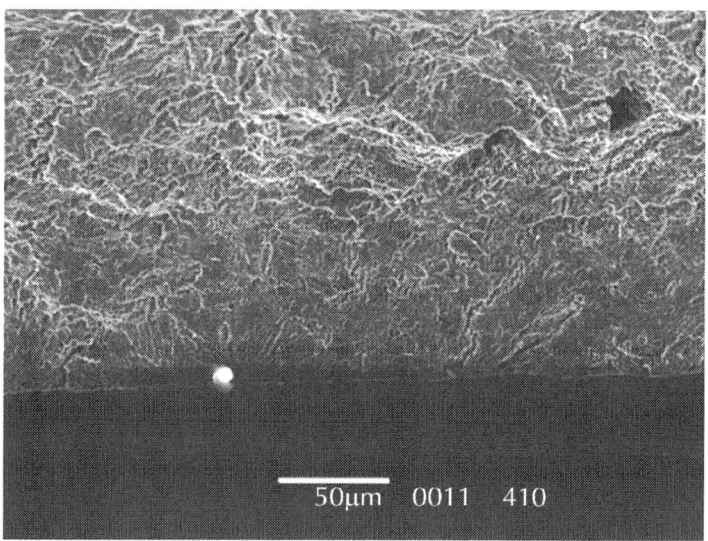

Beach marks
of subordinate
fatigue area

50μm 0000 410

Figure 7: The beach marks of subordinate fatigue area.

50μm   0011   410

Figure 8: The linear origin of subordinate fatigue fracture surface.

There are a lot of beach marks on the entire fracture surface, especially on the later propagation area. In addition, fatigue striations between the beach marks can be seen in the later propagation area, as shown inFig. 9. Every fatigue beach mark on fatigue fracture surface of the 4th disc corresponded to one engine start or one marked state change of the engine [1], [2] and [3]. The beach marks were measured from origin to the instantaneous area in the main fatigue propagation direction. 134 beach marks have been found. Some distances of the 134 beach marks to origin in propagation direction have been measured, as shown in Fig. 10.

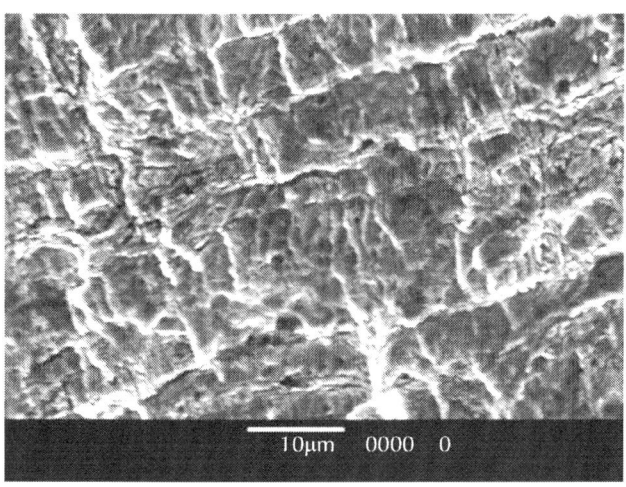

Figure 9: The dense fatigue striation.

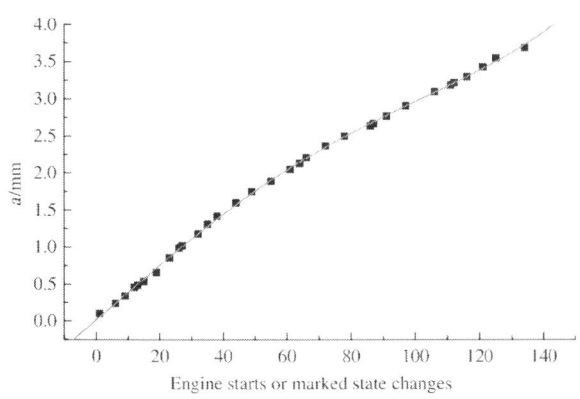

Figure 10: Crack length related to engine starts or marked state changes.

Material Quality Analysis

Metallurgical samples were removed from the web in parallel and vertical directions, respectively. The material of web has dot shape inclusions and micro structure with temper martensite and a little carbide. The microstructure is normal. Hardness samples were removed from the web and the HB hardness meets with the technical requirement. Chemical composition analysis samples were removed from the web. Chemical composition meets with the technical requirement. The web thickness at the fracture section is also measured, and the measurement result indicates that the thickness meets with the design requirement.

Vibration Peculiarity Analysis of the Engine

The number of central transmission main gear is 36, the fatigue stress with the frequency interrelated with n_2 caused by the vibrations is onwards along the axes and was passed to the 4th disc directly. The number of first rotor active blades is 24, the frequency of the vibration stress is correlative to n_1, and the fatigue stress act on the blades caused by the vibrations is onwards along the axes and was passed to the 4th disc indirectly. The two types of vibration stress are main vibration fountain, but not dangerous to the engine.

Traveling wave vibration is caused by the relative running between the nodes radius line on the disc and the disc. The nodes radius type vibration is usually traveling wave model. The fatigue stress act on the disc will induce traveling wave. The constant unilateralism stress with the direction of verticality to the disc will lead to standing wave. The traveling wave and standing wave vibration on the disc both have certain fatalness.

Laboratory tests and computer simulation results show that there would happen a high-order multiple vibration with a narrow frequency band of 6550 Hz when the engine rotation speed is at n_1 = 96 ~ 97% and n_2 = 97 ~ 98%. The vibration is a type of 2 nodes in circumference and 12 nodes in radius, as shown in Fig. 11. The biggest stress for this vibration is located at the R3 arc transition line. The biggest radius vibration stress is between 72 MPa and 81 MPa, centrifugal stress is about 370 Mpa. The yield strength of the disc material is not lower than 885 MPa and the fatigue limit is about 490 MPa at 300 °C, which shows that the counter fatigue capability of the flange is low, and the fatigue damage easily happen under resonance state. Especially, when there are corrosion pits at R3 arc transition, the fatigue crack maybe occurs at the R3 arc between the web and the flange of the 4th disc. The 4th disc can be used in normal condition, however, under certain circumstances, the disc may fail at the R3 arc transition.

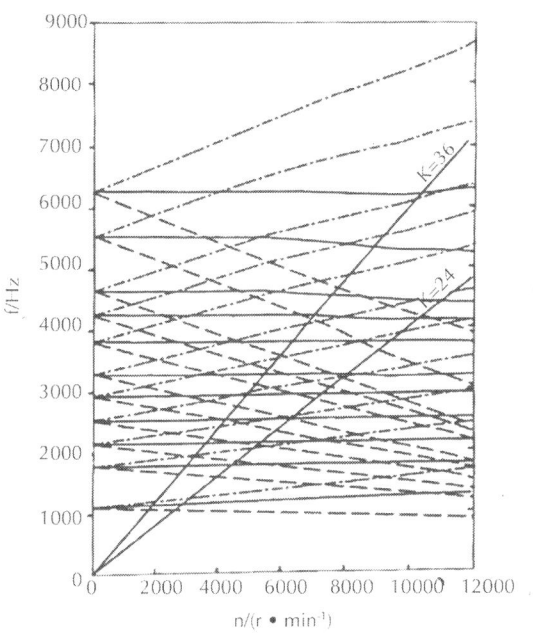

Figure 11: The chart of traveling wave vibration with S = 2 and N = 12.

DISCUSSION AND ANALYSIS

The fracture of the 4th compression disc is high cycle fatigue. The fatigue failure is related to the traveling wave vibration under certain engine rotation speeds and corrosion pits at the R3 arc transition line between the web and the flange.

The main fatigue crack originated from front side surface and propagated to the rear side surface of the web. The fracture surface is flat, exists at least 134 beach marks and fine fatigue striations.

The fatigue origin, with fine radiations and without fatigue steps, is also flat. All of these characteristics are typical features of high cycle fatigue, which implies that the stress changes in propagation stage were not remarkable and the fatigue crack has existed for a long time, i.e., the crack mainly propagated by relatively low vibration stress. Without typical corrosion fatigue characteristics, the propagation area additionally proves that the crack propagated under high frequency and low vibration stress [4] and [5].

There existed corrosion pits on the disc web, especially on the R3 transitional arc. The main crack initiated at the corrosion pit on R3 arc transition line. Corrosion pits were resulted from oceanic climate that contains much of chlorine ion. The chlorine ion in air may badly erode the 4th disc made from 1Cr11Ni2W2MoV and may generate a mass of corrosion pits at the R3 arc transition. The stress at R3 arc transition between the web and the flange are higher in engine operation. Additionally, the corrosion pits at this position damaged the surface integrity, increased the stress because of stress concentration, and resulted in the early fatigue failure.

Originated from the corrosion pit, the main crack extended to the two-sides. However, on the right side, crack growth stopped due to the quick departure from the R3 arc transition line. While on the left side, the crack propagated along the R3 arc transition line for a certain length, therefore, a fatigue crack with big area in the left and small area in the right was formed.

The subordinate fatigue area is small and near the later propagation area, and originated from the machining marks and propagated to the front side surface, which indicate that the stress increased greatly at the arc transition in the later propagation stage, and that axial vibration was generated on the 4th disc.

CONCLUSIONS

- The fracture mechanism of the 4th compressor disc is high cycle fatigue.
- The main fatigue region of the 4th disc is located at the R3 arc transition between the web and the flange, and the fatigue originated from the corrosion pit at the front side surface of the web.

- The microstructure, hardness, chemical composition and the web thickness of the fracture site all meet with technical requirements. The fatigue failure has no connection with the material quality.

- The corrosion pits at the R3 arc transition between the web and the flange induced accident. The cracking and propagation of the 4th compression disc is mainly related to the vibration stresses.

REFERENCES

1. Zhang Zheng, Zhang Haiying, Tian Yongjiang, Zhong Qunpeng. Fatigue stress estimation by fracture surface characteristics. China Mech Eng 2001;12(3):337–9.

2. Liu Xinling, Zhong Oeidao, Zhang Weifang. Fracture analysis on the 3th turbine blade of some engine. Mater Mech Eng 2005;29(8):67–70.

3. Zhang Yi, Zhang Wenfeng, Yan Hai. The function of fractography quantitative analysis during the life estimation of component. J Mater Eng 2000(4):45–8.

4. Zhang Dong, Zhong Peidao, Tao Chunhu. Failure analysis. Beijing: National Defence Industry Press; 2004.

5. Tao Chunhu, Zhong Peidao, Wang Renzhi, Nie Jingxu. Failure analysis and prevention for rotor in aero-engine. Beijing: National Defence Industry Press; 2000.

Chapter 8

Metal Hydride Hydrogen Compressors: A Review

M.V. Lototskyy[a], V.A. Yartys[b, c], B.G. Pollet[a],
and R.C. Bowman Jr.[d]

[a]HySA Systems Competence Centre, South African Institute for Advanced Materials Chemistry, University of the Western Cape, Private Bag X17, Bellville 7535, South Africa

[b]Institute for Energy Technology, Kjeller NO-2027, Norway

[c]Norwegian University of Science and Technology, Trondheim NO-7491, Norway

[d]Oak Ridge National Laboratory, Oak Ridge, TN 37831, USA

ABSTRACT

Metal hydride (MH) thermal sorption compression is an efficient and reliable method allowing a conversion of energy from heat into a compressed hydrogen gas. The most important component of such a thermal engine – the metal hydride material itself – should possess several material features in order to achieve an efficient performance in the hydrogen compression. Apart from the hydrogen storage characteristics important for every solid H storage material (e.g. gravimetric and volumetric efficiency of H storage, hydrogen sorption kinetics and effective thermal conductivity), the thermodynamics of the metal–hydrogen systems is of primary importance resulting in a temperature dependence of the absorption/desorption pressures). Several specific features should be optimised to govern the performance of the MH-compressors including synchronisation of the pressure plateaus for multi-stage compressors, reduction of slope of the isotherms and hysteresis, increase of cycling stability and life time, together with challenges in system design associated with volume expansion of the metal matrix during the hydrogenation.

The present review summarises numerous papers and patent literature dealing with MH hydrogen compression technology. The review considers (a) fundamental aspects of materials development with a focus on structure and phase equilibria in the metal–hydrogen systems suitable for the hydrogen compression; and (b) applied aspects, including their consideration from the applied thermodynamic viewpoint, system design features and performances of the metal hydride compressors and major applications.

INTRODUCTION

Metal Hydride (MH) hydrogen compression utilises a reversible heat-driven interaction of a hydride-forming metal, alloy or intermetallic compound with hydrogen gas to form MH and is considered as a promising application for hydrogen energy systems. This technology, which initially arose in early 1970s, still offers a good alternative to both conventional (mechanical) and newly developed (electrochemical, ionic liquid pistons) methods of hydrogen compression. The advantages of MH compression include simplicity in design and operation, absence of moving parts, compactness, safety and reliability, and the possibility to consume waste industrial heat instead of electricity.

Results of more than 40 years of R&D activities in the development of MH hydrogen compression have been reported in numerous original research papers, patents, reports and conference presentations. However, few review articles on the topic are available. A brief review on the principle of H_2 compression using MH, related R&D within the field and their own feasibility studies of MH H_2 compression was published by Lynch et al. in 1984 [1]. A detailed consideration of the related MH-based thermodynamic engines (heat pumps) was presented by Dantzer and Orgaz in three review papers [2], [3] and [4], 1986–1987. A general approach to the development of the MH hydrogen compressors for various applications based on thermodynamic analysis was considered by Solovey in 1988 [5]. Rather comprehensive reviews of MH compressors and heat pumps were published as sections of general review papers on applications of metal hydrides, by Sandrock in 1994 [6] and Dantzer in 1997 [7]. Bowman has reviewed the development of metal hydride compressors for the liquefaction of hydrogen via the Joule–Thomson process [8] and [9]. Status of the development of metal hydride based heating and cooling systems was summarised in a paper by Muthukumar and Groll [10] in 2010.

The present review summarises the state of the art of the MH hydrogen compression technology, by considering and discussing the relevant data in materials and systems development, analysis of design features and performances of the MH compressors, and their applications. For the sake of better understanding of the processes taking place in the MH hydrogen compressors, the first section of the review presents relevant fundamental aspects focused on the consideration of the suitable hydride forming materials for hydrogen compressors.

METAL–HYDROGEN SYSTEMS FROM A FUNDAMENTAL VIEWPOINT

Applications of metal hydrides, including hydrogen compression, utilise a reversible heat-driven interaction of a hydride-forming metal/alloy, or intermetallic compound (IMC) with hydrogen gas, to form a metal hydride:

$$M(s) + x/2\,H_2(g) \xrightarrow[\text{desorption}]{\text{absorption}} MH_x(s) + Q:$$

(1)

where M is a metal/alloy (e.g., V or a BCC solid solution based upon it), or an IMC ($LaNi_5$, TiFe, etc.); (s) and (g) relate to the solid and gas phases, respectively. The direct interaction, an exothermic formation of the metal hydride/hydrogen absorption, is accompanied by a release of heat, Q. The reverse process, endothermic hydride decomposition/hydrogen desorption, requires supply of approximately the same amount of heat.

The following gas phase applications of metal hydrides use specific features of the Reaction (1)[6], [7], [8],[9], [10], [11] and [12]:

- Compact and efficient hydrogen storage is due to a very high, about 100 g_H/L, volumetric density of atomic hydrogen accommodated in the crystal structure of the MH metal matrix. At ambient temperatures the equilibrium of the Reaction (1) can often take place at modest, ≤1–10 bar hydrogen pressures. Thus, hydrogen storage using MH is intrinsically safe and benefits from avoiding use of compressed hydrogen gas and energy inefficient and potentially unsafe liquid H_2. Endothermic reverse process of dehydrogenation according to the Reaction (1) decreases temperature of the MH leading to decreased rates of H_2 evolution; this, in turn, is an additional safety feature of use of the MH, allowing to avoid accidents even in case of rupture of the hydrogen storage containment.

- Simple and efficient pressure/temperature swing absorption–desorption systems. This allows not only to control hydrogen pressure by changing temperature, but, furthermore opens possibilities forhydrogen separation and purification (including isotope separation) due to the high selectivity of theReaction (1).

- Reversibility and significant heat effects (≥20 kJ/mol H_2) of the Reaction (1) make it possible to realise numerous energy conversion applications of MH. This includes first of all thermally driven hydrogen compression and heat management.

- The process performances, especially for the latter applications considered in the present review, are strongly dependent on the intrinsic features of the Reaction (1) including its thermodynamic and kinetic characteristics (the macro-kinetic parameters involving heat-and-mass transfer issues are also very important), as well as composition, structure and morphology of the solid phases (M, MHx) involved in the process. These features, mainly related to fundamental aspects of MH materials science, are considered in the current section.

Phase Equilibria in the Metal–Hydrogen Systems

Equilibrium of the Reaction (1) is characterised by an interrelation between hydrogen pressure (P), concentration of hydrogen in the solid phase (C) and temperature (T). This relation (PCT-diagram) is the characteristic feature of a specific hydride-forming material determining thermodynamics of its interaction with gaseous hydrogen. At the same time, thermodynamic behaviour of the metal–hydrogen systems has common characteristics, which are similar for different materials [13].

At low hydrogen concentrations $(0 \leq C < a$ $)$ hydrogen atoms form an interstitial solid solution in the metal matrix (-phase) with

$C(H) \sim \sqrt{P(H_2)}$ according to a Henry–Sieverts law. When the value of C exceeds concentration of the saturated solid solution (a), precipitation of the hydride (β-phase with hydrogen concentration $b > a$) occurs, and the system exhibits features of first order phase transition taking place at a constant hydrogen pressure, $P = PP$ $(a \leq C \leq b)$. This pressure is called as plateau pressure in the diagrams of the metal–hydrogen systems. Further increase in hydrogen concentration is again accompanied by the pressure increase corresponding to the formation of H solid solution in the β-phase. When the concentration approaches a certain maximum value $(C \rightarrow C_{max})$ corresponding to the maximum hydrogen storage capacity of the material, or the number of interstitial sites available for the insertion of H atoms, the equilibrium pressure exhibits an asymptotic increase, $P \rightarrow \infty$.

The plateau width, $(b–a)$, is often considered as a reversible hydrogen capacity of the material, and the equilibrium of Reaction (1) in the plateau region is described by van't Hoff equation:

$$\ln \left(\frac{P_P}{P^0} \right) = -\frac{\Delta S^0}{R} + \frac{\Delta H^0}{RT};$$

(2)

where $P^0 = 1$ atm $= 1.013$ bar, [1] ΔS^0 and ΔH^0 are the standard entropy and enthalpy of hydride formation respectively, R is the gas constant.

The values of plateau pressures, PP, at a given temperature are thus dependent on ΔS^0 and ΔH^0 which are individual properties of the material. For various hydride forming alloys and IMC's, ΔS^0 varies insignificantly around -111 ± 14 J/(mol H_2 K), see Table 1; that value is close for different systems as this is the change of entropy of gaseous H_2 during the Reaction (1) originating from the main/configurational contribution (about -130 J/(mol H_2 K)) to the entropy from dissociation of H_2. Consequently, the plateau pressure will be mainly determined by the reaction enthalpy, H^0, which widely varies for different metals and is a measure of the average strength of the M–H bond in MHx [14]. The latter is strongly dependent on the composition and crystal structure of the parent metallic material, including type of its components (as regards to their affinity to hydrogen), their stoichiometry and interaction energy in the alloy or IMC, type/surrounding and size of the interstitial sites in the metal matrix available for the insertion of the H atoms.

Table 1: Equilibrium characteristics of the interaction of hydride-forming alloys suitable for H_2 gas in plateau region The data are sorted in the ascending order for desorption plateau pressure at $T = 25$ °C (P_0). The plateau pressures are calculated using Equation (2); the lower (P_L) and higher (PH) values correspond to the lower (TL) and higher (T_H) temperatures, respectively, as reported in the original works

#[a]	Alloy	−S° [J/(mol H_2 K)]	−H° [kJ/mol H_2]	Temperature range [°C]		Pressure [atm]			Ref.
				TL	TH	P_0	PL	PH	
1 (A)	$V_{75}Ti_{12.5}Zr_{12.5}$	145.1	52.98	30	120	0.02	0.03	3.47	[15]
2 (B)	$MmNi_{4.8}Al_{0.2}$	111.3	37.20	50	150	0.02	0.63	16.66	[16][b]
3 (B)	$LaNi_{4.7}Sn_{0.3}$	112.6	36.51	25	80	0.31	0.31	3.03	[17]
4 (A)	$V_{75}Ti_{10}Zr_{7.5}Cr_{7.5}$	132.3	42.23	30	120	0.32	0.43	19.90	[15]
5 (B)	$LaNi_{4.8}Sn_{0.2}$	104.3	32.83	20	90	0.50	0.40	5.32	[18][b]
6 (B)	$Mm_{0.5}La_{0.5}Ni_{4.7}Sn_{0.3}$	105.0	32.80	0	240	0.55	0.16	139.9	[19]
7 (B)	$LaNi_{4.8}Al_{0.2}$	111.2	33.80	25	80	0.77	0.77	6.44	[17]
8 (B)	$LaNi_5$	101.6	30.40	50	150	0.96	2.47	35.84	[16][b]
		110.0	31.80	25	200	1.49	1.49	171.9	[20] and [21][b]
9 (B)	$MmNi_{4.7}Fe_{0.3}$	87.4	25.00	20	102	1.53	1.29	12.14	[22][b]
10 (A)	$V_{0.85}Ti_{0.1}Fe_{0.05}$	148.0	42.90	−20	100	1.64	0.08	53.14	[20] and [23][b]
11 (C)	$TiFe_{0.9}Mn_{0.1}$	107.7	29.70	0	100	2.64	0.88	29.39	[20] and [24][b]

12 (B)	$La_{0.85}Ce_{0.15}Ni_5$	91.28	24.30	10	110	3.24	1.93	28.50	b,c
13 (B)	$MmNi_{4.7}Al_{0.3}$	107.8	28.88	20	90	3.73	3.05	29.98	[18]
14 (A)	$V_{92.5}Zr_{7.5}$	147.0	40.32	30	60	4.11	5.38	22.71	[15]
15 (B)	$La_{0.2}Y_{0.8}Ni_{4.6}Mn_{0.4}$	105.3	27.10	20	90	5.62	4.67	39.78	[25]
16 (D)	$Zr_{0.7}Ti_{0.3}Mn_2^d$	85.0	21.00	30	150	5.77	6.63	70.41	[26][b]
17 (D)	$Ti_{0.9}Zr_{0.1}Mn_{1.4}Cr_{0.35}V_{0.2}Fe_{0.05}^c$	106.9	25.89	2.5	100	11.17	11.17	91.14	[27]
18 (B)	$MmNi_{4.15}Fe_{0.85}$	105.4	25.00	2.5	200	11.36	11.36	502.8	[20] and [24][b]
19 (B)	$La_{0.6}Ce_{0.4}Ca_{0.2}Ni_5$	115.3	28.20	15	100	12.08	8.14	118.9	[28]
20 (D)	$Ti_{0.8}Zr_{0.2}CrMn$	108.6	24.60	−20	50	23.06	3.95	49.69	[20][b]
21 (B)	$Mm_{1-x}Ca_xNi_{5-y}Al_y^c$	103.0	22.85	5	90	23.82	12.28	124.0	[29]
22 (B)	$Ca_{0.2}Mm_{0.8}Ni_5$	109.5	24.50	0	100	26.75	10.83	195.0	[20] and [24][b]
23 (D)	$Zr_{0.8}Ti_{0.2}FeNi_{0.8}V_{0.2}$	118.3	26.80	20	90	30.49	25.35	211.1	[30][b]
24 (D)	$Ti_{0.77}Zr_{0.3}Cr_{0.85}Fe_{0.7}Mn_{0.25}Ni_{0.2}Cu_{0.01}$	93.66	19.26	20	110	32.98	28.88	184.8	c
25 (D)	$TiCr_{1.9}Mo_{0.01}$	113.0	24.80	−50	90	36.11	1.25	216.4	[30]
26 (D)	$TiCr_{1.9}$	122.0	26.19	−100	30	60.77	0.03	72.34	[31][b]
27 (D)	$ZrFe_{1.8}Cr_{0.2}$	109.0	22.30	20	90	61.19	52.49	306.2	[30]

28 (D)	$(Ti_{0.9}Zr_{0.03})_{1.1}Cr_{1.6}Mn_{0.4}$	115.0	23.40	10	99	80.80	49.00	527.9	[32]
29 (D)	$TiCr_{1.5}Mn_{0.25}Fe_{0.25}$ᵃ	101.6	19.32	−10	165	83.61	29.65	1009	[27]
30 (D)	$TiCr_{1.5}Mn_{0.2}Fe_{0.1}$ᵃ	101.0	18.32	−10	148	116.4	43.57	1008	[27]
31 (D)	TiCrMn	106.0	19.60	−60	100	126.8	5.42	621.2	[34]ᵇ
32 (D)	$ZrFe_{1.8}Ni_{0.2}$	119.7	21.50	20	90	306.0	264.0	1445	[30]
33 (D)	$Ti_{0.86}Mo_{0.14}Cr_{1.9}$	117.0	17.20	−50	90	1253	121.7	4340	[30]

[a]Type of the alloy is specified in brackets as BCC-V solid solution (A); AB_5- (B), AB- (C) and AB_2-type (D) IMC's.

[b]The data are also available at the US DoE hydrogen storage materials database, http://hydrogenmaterialssearch.govtools.us; section "Hydride Information Center (Hydpark)".

[c]Previously unpublished experimental data by the authors of this review (ML, VY).

[d]Dynamic PCT experiments.

[e]ΔS^0 fitted by ML to agree with the reported T–P conditions.

Since *PP* increases exponentially with temperature, the low-temperature H absorption at $PH_2 > PP(TL) = PL$ takes place at a lower hydrogen pressure, and the high-temperature H desorption ($PH_2 < PP(TH) = PH$) occurs at a higher pressure, similar to the suction and discharge processes in a mechanical compressor.

Table 1 presents the equilibrium properties of hydrogen interaction with some hydride-forming alloys and IMC's suitable for hydrogen compression applications. Van't Hoff plots for some of these materials are presented in Fig. 1; typical requirements for the H_2 compression (P = 1–400 atm, T = 25–150 °C) are shown as a rectangular area.

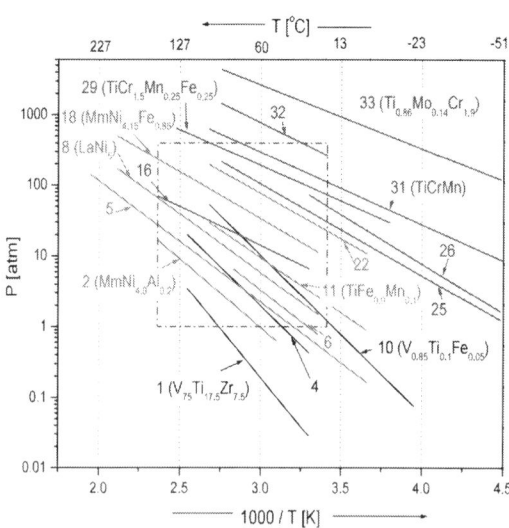

Figure 1: Van't Hoff plots for selected hydride-forming alloys suitable for H_2 compression. Plot numbers correspond to the numbers of alloys in Table 1;

plot colours correspond to the types of the hydride-forming alloys (A – black, B – red, C – olive, D – blue). Rectangular area limited by dash-dot line shows target requirements for H_2 compression: P = 1–400 atm, T = 25–150 °C. (For interpretation of the references to colour in this figure legend, the reader is referred to the web version of this article.)

It can be seen that, depending of the type (A–D) and composition of the hydride-forming material, the equilibrium hydrogen pressures vary in a very broad range, from below 1 bar to exceeding 1 kbar at room temperature. Most of the lower-pressure H_2 compression alloys (PH < 200 bar at TH < 150 °C) belong to the AB_5-type intermetallic compounds (group B in Table 1) while significantly higher, >1 kbar, hydrogen pressures can be generated using AB_2-type IMC's (group D).

As it can be seen from Fig. 2, hydrogen compression ratio (PH/PL) achieved using MH in the temperature range from TL ~ 25 °C to TH = 100–150 °C varies in the range 10–50 at PH = 100 atm. The value of PH/PL has a tendency to become smaller when PH increases, but remains quite high (5–10) even for the H_2 discharge pressures ≥1 kbar.

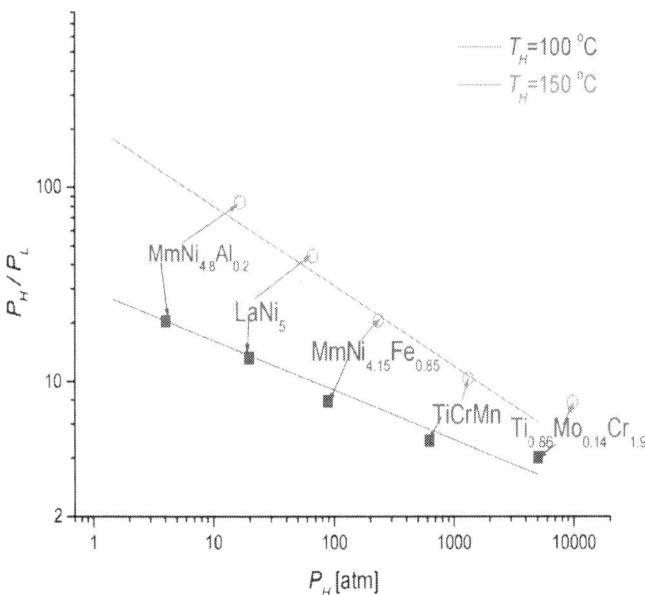

Figure 2: Dependencies of hydrogen compression ratio at TL = 25 °C for selected hydride-forming alloys (Table 1).

It has to be noted that the presented above hydrogen compression performances calculated on the basis of van›t Hoff Equation (2) are only rough estimates which significantly deviate from real characteristics of metal hydride materials, even being considered only from thermodynamic point of view.

The major factor affecting hydrogen compression efficiency of the MH materials is the plateau slope. In a multi-component hydride-forming IMC›s (e.g., ABn) the sloping plateaux are originated from compositional fluctuations due to the presence of impurities randomly substituting A- and/or B-component, or because of fluctuations of the stoichiometry (AB$n_{\pm}x$) within the homogeneity region [35]. The quantification of this phenomenon by introducing statistical (as a rule, Gaussian) distribution of PP was first suggested by Larsen and Livesay [36] and further developed by Fujitani et al. [37], Lototsky, Yartys et al. [38] and [39], Park et al. [40].

In addition to operating pressure–temperature ranges, an important parameter of MH material for hydrogen compression is the process productivity. The simplest approach for its estimation assumes the productivity of H$_2$ compression cycle (per unit of weight or per number of the metal atoms) as (b–a), i.e. plateau width, where the values of b and a are available. The problem of this approach is that both a and b are temperature-dependent, and the plateau width decreases with increase of the temperature. Furthermore, at a critical temperature, TC, the plateau degenerates to an inflection point, and at $T > TC$, the pressure–composition isotherms are continually sloping [14]. Hence, for a realistic estimation of the hydrogen compression productivity it is necessary to know temperature dependencies of a and b, or to have a quantitative information about phase diagram of the hydrogen–metal system. The corresponding approach for the modelling of PCT diagrams using statistical and thermodynamic features was suggested by Lacher for H–Pd system already in 1937 [41] and further developed by Kierstead [42], Brodowsky et al. [33], Beeri et al. [34], Lototsky, Yartys et al. [39]. Finally, hydrogen compression performances of the real MH systems are significantly affected by hysteresis, as the values of plateau pressures for hydrogen absorption/hydrogenation are higher than the ones for hydrogen desorption/dehydrogenation. Hysteresis is caused by stresses which appear in the course of growth of MH nuclei inside the matrix of the MH alloy having lower molar volume. The thermodynamic aspects of hysteresis were discussed in detail in a number of publications (see, e.g. Refs.[13], [43] and [44]). The influence of hysteresis on the performance of MH hydrogen compressors will be discussed in section 3.2. Taking into account the features of phase equilibria in the real metal–hydrogen systems described above, we can illustrate the process of thermally-driven hydrogen compression using MH by the scheme shown inFig. 3.[2] Hydrogen is absorbed in the MH at a lower temperature, TL, following the hydrogen absorption isotherm at TL (1); the process is accompanied by a release of heat, Q $\approx |\Delta H^0|$. The absorption is carried out at a lower pressure, so the system approaches equilibrium which corresponds to the point B on the isotherm (1). The corresponding value of hydrogen concentration (CL) is strongly dependent on the hydrogen pressure and, generally, it is not equal to the lower limit, $b(1)$ (see Fig. 3(a)), of hydrogen concentration in β-hydride at TL.

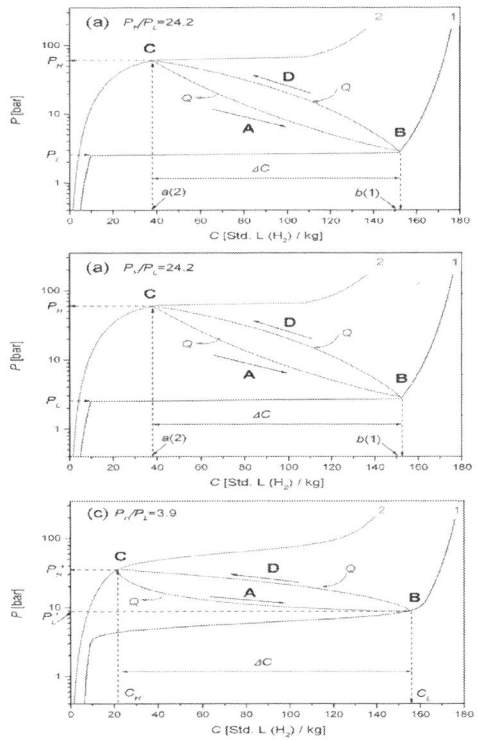

Figure 3: Pressure–composition isotherms at $TL = 20\ °C$ (1) and $TH = 150\ °C$ (2) for H–La$_{0.85}$Ce$_{0.15}$Ni$_5$ system illustrating thermally-driven hydrogen com-

pression using MH: (a) – idealised (flat plateaux, desorption isotherms), (b) – idealised (sloping plateaux, desorption isotherms), (c) – real (sloping plateaux, absorption isotherm at TL, desorption isotherm at TH).

Further heating of the system to a higher temperature, TH, results in the hydrogen desorption from MH which follows the hydrogen desorption isotherm at TH (2) and requires absorption of heat, Q. When the desorbed hydrogen is released at a higher pressure, the system equilibrium corresponds to the point C on the desorption isotherm (2). Similarly, hydrogen concentration (CH) in this point depends on the hydrogen desorption pressure and may not be equal to hydrogen concentration, $a(2)$ (see Fig. 3(a)), in the saturated α-solid solution at TH.

In the real systems, due to sloping plateau (Fig. 3(b)) and hysteresis (Fig. 3(c)), hydrogen compression in the same temperature range (from TL to TH corresponding to isotherms 1 and 2, respectively) will require higher suction pressures ($PL'>PL$) and lower discharge pressures ($PH' < PH$) than the corresponding values calculated by Van't Hoff Equation (2) using ΔS^0 and ΔH^0 reference data (Table 1; usually provided for desorption). Accordingly, the compression ratio, PH'/PL', will be lower than the PH/PL estimation based on the ideal plateau behaviour, Fig. 3(a).

Hydrogen compression from PL' to PH' carried out in the temperature range from TL to TH is represented by a cyclic process involving hydrogen absorption (A) and hydrogen desorption (D) between points B and C on the isotherms 1 (H absorption) and 2 (H desorption), respectively.

Independent of specific paths B → C (D) and C → B (A), the amount of hydrogen taking part in the compression cycle (or cycle productivity of the process) will be equal to the change of hydrogen concentration in the solid (ΔC). For the specific MH material this value will be strongly dependent on the process conditions (TL, TH, PL' and PH'; Fig. 3(c)). The described evaluations are based on maintaining of both TL and TH, which requires enhancements of the heat transfer [45] due to the poor thermal conductivity of hydride powders (see 2.2 and 3.3).

Fig. 4 presents calculated values of ΔC for $La_{0.85}Ce_{0.15}Ni_5$ at $TL = 20$ °C and $TH = 150$ °C. As it can be seen, at the suction pressure (PL') above the midpoint of the sloping plateau of the H absorption isotherm, the hydrogen compression productivity significantly increases.

Increase of the discharge pressure (PH') results in the significant loss of the productivity. However, if the suction pressure is high enough (that corresponds to β-region of the H absorption isotherm (1), see Fig. 3), very high discharge pressures can be generated with the productivity about 20% of the reversible hydrogen capacity of the material at TL. This effect has its origin in (i) decrease of the lower limit, b, of hydrogen concentration in β-hydride with increasing the temperature, and (ii) contribution of the dissolved hydrogen additionally released from the β-hydride. Increase of TH will result in further increase of the discharge pressure with no significant changes in productivity, due to the lowering of the concentration, a, of the saturated α-solid solution.

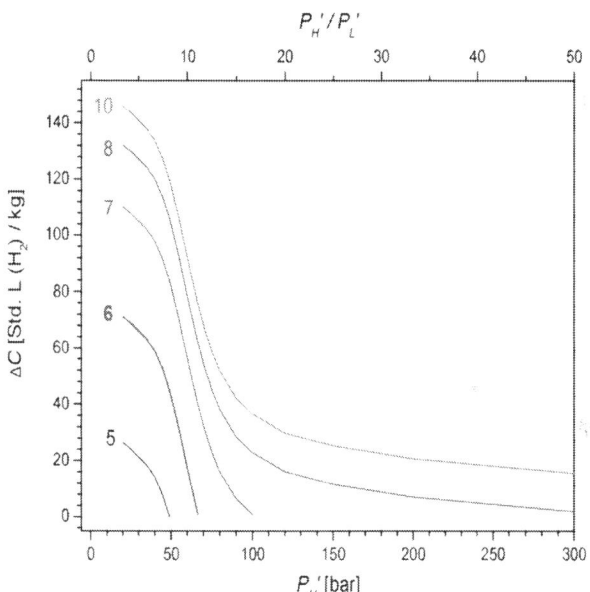

Figure 4: Dependence of H_2 compression cycle productivity, ΔC, for the H–$La_{0.85}Ce_{0.15}Ni_5$ system (Fig. 3) on the H_2 desorption pressure,PH', at $TH = 150$ °C. Curve numbering corresponds to the H_2 absorption pressure, PL' [bar], at $TL = 20$ °C. The value $PL' = 6$ bar corresponds to the plateau midpoint.

The feasibility of generating high H_2 pressures using quite stable MH was mentioned by Golubkov and Yuhimchuk [46] who reported about compression of hydrogen isotopes to 57 bar ($TiH_2/TH = 700$ °C), 85 bar ($UH_3/700$ °C) and 700 bar ($VH_2/250$ °C).

In summary, selection of the MH materials able to provide required H_2 compression from *PL* to *PH* in the available temperature range (*TL* to *TH*) can be achieved by analysing dependence of thermodynamic properties (enthalpy and entropy) of hydrogenation/dehydrogenation, on the alloy composition. More accurate thermodynamic estimations of hydrogen compression performances of MH materials, including suction (*PL'*) and discharge (*PH'*) pressures at cooling (*TL*) and heating (*TH*) temperatures, compression ratio (*PH'/PL'*) and the process cycle productivity (ΔC), are possible when considering the complete isotherms of hydrogen absorption at *TL* and hydrogen desorption at *TH*, taking into account plateau slope, hysteresis and features of H–M phase diagram. This assessment can be done by the fitting the available experimental PCT data using suitable models of phase equilibria in metal–hydrogen systems.

H Sorption/Desorption Kinetics

Though cycle productivity of H_2 compression using MH can be determined from the thermodynamic considerations (see previous section), the duration of the hydrogen absorption–desorption cycle and, correspondingly, dynamic performances of MH hydrogen compressors depend on the rates of the direct and reverse processes of Reaction (1), i.e. on kinetics of hydrogen absorption and desorption which can vary significantly from alloy to alloy. Because many intermetallic hydrides exhibit rather fast intrinsic hydrogenation–dehydrogenation kinetics, the rates of H absorption and desorption for the MH storage materials are generally more often limited by heat transfer [45] and [47]. In some cases, e.g. operation at low temperatures, or in presence of gaseous impurities in hydrogen gas, kinetic factors may become decisive [48]. Thus, heat-and-mass transfer modelling of H_2 charge/discharge flow rates in the MH should incorporate a reliable and verified kinetic expression of the rates of H uptake and release [49] and [50].

Kinetics and mechanism of hydrogen–metal interaction were analysed in numerous original research papers and review publications (see, e.g. Refs. [49] and [51]). The interrelation between kinetics and heat transfer determining the eventual rates of H_2 absorption and desorption in various reactors was first considered by Goodell [47]. He concluded that, due to the very fast isothermal H_2 absorption (estimated time of 75% hydrogenation for LaNi$_5$ at PH$_2$ = $2PP$ and $T =$

25 °C equals to just 0.5 s) and poor effective thermal conductivity of the MH powder (~1 W/(m K)), the system is quickly self-heated and approaches equilibrium elevated temperature conditions; these can be calculated by solving the van't Hoff Equation (2)taking plateau pressure, PP, as the actual H_2 pressure. Further H_2 absorption or desorption is limited by the rate of cooling or heating of the MH. Consequently, the most important kinetic aspect in the dynamic behaviour of the MH reactors is hydrogen absorption–desorption rate at near-equilibrium conditions. As it was shown by Førde, Yartys et al. in Ref. [49], a good approximation of the reaction kinetics in this case can be achieved by using Avrami–Erofeev equation:

$$X = 1 - \exp[-n(Krt)]; \tag{3}$$

where X is the reacted fraction, Kr is the rate constant, t is time, and n is integer or half-integer whose value (0.5...4) depends on the reaction mechanism. The reaction fraction is defined as:

$$X \equiv \frac{C_H - C_1}{C_2 - C_1}; \tag{4}$$

where CH is the actual hydrogen concentration; C_1 and C_2 is the hydrogen concentrations at the beginning and end of the reaction, respectively.

The value of the rate constant can be presented as a product of the pressure-defined driving force of the process, $K(P)$, and Arrhenius-like pressure independent term:

$$K_r = K(P) \cdot K_0 \exp\left(-\frac{E_a}{RT}\right) \tag{5}$$

where Ea is activation energy.

Note that the pressure driving force, $K(P)$, depends on the deviation of the actual hydrogen pressure from the equilibrium one (typical dependencies for the various reaction mechanisms, e.g. $K(P) = \ln(P_{eq}/P)$ for the desorption, are reviewed in Ref. [49]), and the reaction rates at the given pressure–temperature conditions will be dependent on both kinetic parameters and PCT characteristics of the hydrogen–metal system. This approach is used in the heat-and-mass transfer modelling of the MH reactors for hydrogen compression (section 3.3).

Materials Challenges and Their Solution

Hydrogen compression applications pose the following requirement to MH materials [11] and [48]:

- Tuneable PCT properties allowing to achieve required hydrogen compression ratio (*PL* to *PH*) in the available temperature range (*TL* to *TH*);
- High reversible H storage capacity to minimise the amount of MH and to reduce the energy consumption and the heat losses associated with thermal swings;
- Fast kinetics of hydrogen exchange to achieve higher productivities;
- Low plateau slope of the H absorption and desorption isotherms;
- Low hysteresis, *PA/PD*;
- Cycle stability when operating at high temperatures and H_2 pressures;
- Tolerance of H sorption performances to the impurities in H_2;
- Scaleability of the synthesis of MH alloys and their hydrides, and affordable costs.

The following section briefly describes challenges appearing in the course of the development of MH materials for hydrogen compression and reviews the possible ways of their solution.

Tuning of the Thermodynamic Properties

As it was shown in section 2.1, hydrides of the alloys and IMC's form and decompose in a broad range of equilibrium decomposition pressures. Taking into account the non-ideal behaviours reflected in the shape of the pressure–composition isotherms (primarily, plateau slope and hysteresis), the achievable compression ratio at a reasonable cycle productivity/reversible H capacity is low and seldom exceeds 5–10 at $(TH - TL) \approx 100$ K, Thus, a multistage compression (see section 3.1) is required to reach higher eventual compression values. The multistage operation approach introduces more strict requirements to the tuneability of the PCT characteristics, since in this case the H desorption isotherm at TH for the previous stage and H absorption isotherm at TL for the next stage must be synchronised. The problem of coupling of the MH materials used in the consecutive hydrogen compression stages resembles selection of "high-temperature" and "low-temperature" MH for the heat management applications [3] and [10]. However, in case of hydrogen compression, special attention has to be paid to the operating pressures, in addition to the thermal properties of the corresponding systems.

Altering of the hydrides stability can be achieved by the variation of the composition of the parent alloys. The existent hydride-forming alloys allow very broad, from −70 to −20 kJ/mol H_2, variation in H^0 that corresponds to the H_2 plateau pressures from millibars to kilobars at room temperature [12]. As applied to the commonly used types of hydride-forming alloys and IMC's, the variation in composition offers the following opportunities described in Table 1 and in Fig. 1.

AB5-Type Intermetallics, the most rugged materials for the MH applications, allow variation of the lower/suction pressures from <1 to 20–30 bar at TL = 25 °C, and the higher/discharge pressures from 15–20 to ~200 bar at TH = 100–150 °C. The variations of the thermodynamic stability of the AB_5-based hydrides can be achieved by substitution of lanthanum in $LaNi_5$ by cerium or mischmetal [3] (this lowers the stability and increases the H_2 dissociation pressures), and by substitution of nickel with cobalt, aluminium, manganese, or tin (increasing the stability, and decreasing the H_2 pressures). The AB_5-type IMC's have rather small plateau slope and hysteresis, however, increasing with the increase in cerium/mischmetal, and/or aluminium content [16] and [52]. Introduction of aluminium also results in decrease of the reversible hydrogen capacity [16]. However, Al substitution significantly enhances the durability of the hydride phase during extended absorption/desorption cycling as has been demonstrated for several AB_5-type alloys [53],[54], [55] and [56].

Industrial-scale manufacturing of AB_5-type alloys for hydrogen compression, as well as influence of the substituting components (Ce, Co, Al) in $LaNi_5$ on the operating performances of the MH materials has been studied by Baichtok et al. [57]. The application of $(La,Ce)Ni_5$ for industrial-scale hydrogen compression was reported by Bocharnikov et al. [58].

AB2-Type Intermetallics cover much broader range of the operating pressures. The most stable $ZrV_2H_{\sim 4}$ is characterised by thermodynamic parameters of β–α transition as $\Delta S^0 = -88.4$ J/(mol H_2 K) and $\Delta H^0 = -78$ kJ/mol H_2 [59] that corresponds to plateau pressure below 10^{-6} mbar at room temperature and just around 3 mbar at $T = 300$ °C. At the same time, hydride of $ZrFe_2$ ($\Delta S^0 = -121$ J/(mol H_2 K); $\Delta H^0 = -21.3$ kJ/mol H_2) has plateau pressure above 300 bar at room temperature. A 20% substitution of Zr by Ti results in further destabilisation of the hydride doubling the plateau pressure (this hydride has, however, a huge hysteresis between the pressures of H absorption and desorption) [30]. Since calculated hydrogenation enthalpy of $TiFe_2$ is around -3.6 kJ/mol H_2 [60] and assuming $\Delta S^0 = -100$ J/(mol H_2 K), this will result in a plateau pressure about 40 kbar at a room temperature. Consequently, in the AB_2-type IMC's the operation at higher hydrogen pressures can be achieved by increasing Ti/Zr ratio and Fe content on the, correspondingly, A- and B-sites. The lowering of the operating pressures is achieved by introducing such B-elements as V and, in a lesser extent, Mn and Cr. The multicomponent AB_2-type hydrides appear to show rather high plateau slopes and a profound hysteresis, especially, for higher Fe contents [27] and [30]. Furthermore, the AB_2-type alloys are much more sensitive towards poisoning by the traces of active gases, oxygen and water vapour, when present as admixtures in hydrogen gas [48].

Vanadium-Based BCC Solid Solution Alloys form another type of the materials suitable for MH hydrogen compression. Vanadium forms two hydrides [61], VH_{1-x} and VH_{2-x} where the transition between the mono- and dihydride is characterised by a reversible hydrogen storage capacity of about 1.9 wt.% H at near-ambient conditions and has a steep temperature dependence of hydrogen equilibrium pressure, associated with unusually high values of entropy and enthalpy of the formation of vanadium dihydride. Due to this reason, the BCC vanadium alloys are attractive candidates for the MH hydrogen compressors[62] and [63]. Introduction of ≤17.5 at.% of titanium into V alloys allows significant variation of the plateau pressure, approximately from 0.2 to 10 bar at $T = 60$ °C [15]. Similar variations (0.1–20 bar at $T = 80$ °C) were achieved by introduction of 0–7.5 at.% Fe in $(V_{0.9}Ti_{0.1})_{1-x}Fex$ [64]. Use of some V-based alloys allows for H_2 compression from 20 bar ($T = 10$ °C) to 150–200 bar ($T = 150$ °C) with reversible H capacity exceeding 150 cm^3/g STP. Minor additives of Zr (7.5 at.%), together with Ti (0–

17.5 at.%) and 3d transition metals (Cr, Mn, Fe, Co, Ni; up to 7.5 at.%), significantly improve hydrogenation/dehydrogenation kinetics, and the variation of the hydride stability can be achieved by changes in the amount of Ti and the transition metals [15]. The main disadvantage of the usage of BCC-V alloys for H_2 compression is in quite high hysteresis and significant sloping of the H_2 absorption isotherms; increased H_2 absorption plateau pressures lead to significant hysteresis; hysteresis increases during H absorption–desorption cycling [65].

Tife-Based AB-Type Intermetallics can offer advantage of low costs that makes them an attractive option for the applications of MH. The first heat-driven hydrogen compressor was patented in 1970 by Wiswall and Reilly [66] and used TiFe to compress H_2. However, limited possibilities for the element substitution in TiFe does not allow one to easily vary the stability of its hydrides, as compared to the alloys considered earlier in the review. The other drawbacks of TiFe as a hydrogen compression alloy include presence of two plateaux on the pressure–composition isotherms, high hysteresis, difficulties in activation, and sensitivity to the presence of minor impurities of O_2 and H_2O resulting in a strong deterioration of the hydrogen storage performance. Some improvements can be achieved in the course of alloying of TiFe with Mn or V, addition of deoxidisers (RE metals), as well as variations in the procedures of the material preparation and treatment [48] and [67].

Tolerance to the Impurities In H_2

Tolerance of the hydride-forming materials towards impurities in H_2 is a very important property, especially for such "open-ended" MH applications as hydrogen compression. Depending on the alloy–impurity combination, hydrogen storage properties can deteriorate as a result of various types of damages[6] and [48]:

- poisoning: H storage capacity quickly decreases without a concurrent decrease of intrinsic kinetics;
- retardation: slowing down of the kinetics of hydrogen exchange without a loss of ultimate storage capacity;
- corrosion;
- innocuous damage: no surface deterioration takes place, but there can be pseudo-kinetic decreases due to inert gas blanketing.

The mechanism of the deterioration of hydrogen absorption/ desorption performances is determined by interaction of the impurity with MH surface (1–3), as well as by slowing down of the gas diffusion (4). Influence of the gas impurities can be quantified by empirical equations; their numerical constants depend on the specific "MH– impurity" combination and have to be determined experimentally [68]. The most important impurities for MH H_2 compression process are oxygen and water vapour which are present in hydrogen produced by electrolysis, or traces of carbon dioxide and monoxide in hydrogen produced from carbonaceous feedstock. Moderate concentrations of these impurities (except of CO) are normally not a serious concern for the AB_5-type alloys, but for titanium-based AB_2- and AB-type ones impurities cause problems [48] and [67]. Presence of even ppm-scale amounts of CO results in a strong poisoning of even most tolerant to the impurities AB_5-type alloys [67], [68], [69] and [70].

There are several possibilities in addressing the poisoning problem. Importantly, MH poisoned by the admixtures can be reactivated by vacuum heating [6] and [48]. Secondly, addition of deoxidisers can help as well. One example of the MH hydrogen compressor [71] implies usage of TiFe doped by 2 wt.% of mischmetal. An efficient method of eliminating poisoning is in a surface modification of hydride-forming alloys, for example, by chemical treatment with a fluorine-containing aqueous solution [72]. The coating of the MH surface by transition metals, particularly with platinum group metals, also improves poisoning tolerance and facilitates reactivation procedure; this method combined with the fluorination enables operation in CO-contaminated hydrogen [69] and [70]. Reviews on the surface modification techniques increasing the poisoning tolerance of the MH materials were published by Uchida [73] and Lototsky et al.[69]. Introduction of noble metals into MH particles is a part of MH hydrogen storage and compression technologies by Ergenics, Inc. [74]. Surface modification of vanadium, by acidic leaching followed by ball milling with 20 wt.% $LaNi_5$ was also used by Hu et al. [75] for the preparation of the MH material suitable for the second stage of high pressure hydrogen compressor.

Degradation

In addition to the cycle stability issues caused by impurity of hydrogen gas, hydrogen storage capacity can be lost during extended cycling in pure H_2 because of the side hydrogenation process dictated by thermodynamics of the metal–hydrogen interactions. A degradation of the reversible H storage capacity is caused by a disproportionation of the intermetallic alloy to form a stable binary hydride [68]. As example, in case of $LaNi_5$, the reversible hydrogen absorption and desorption reaction:

$$_5LaNi+3H_2 \rightleftarrows _5LaNiH_6 \qquad (6)$$

is less thermodynamically favourable as compared to the irreversible at the same operation conditions disproportionation process:

$$LaNi_5 + H_2 \rightarrow LaH_2 + 5Ni \qquad (7)$$

Some intermediate processes take place between (6) and (7) and include amorphisation, formation of lattice defects and H-trapping sites. For the IMC's enriched with A-component which forms stable binary hydrides (e.g., Y in YNi_2), the disproportionation also results in the formation of intermetallics enriched with the non-hydrided component B (Ni in YNi_5). Because the disproportionation requires diffusion of the metal atoms, it is strongly retarded at lower temperatures, where a reversible formation–decomposition of the intermetallic hydride according to (6) prevails. However, the degradation processes of the type (7) quickly accelerate when temperature and hydrogen pressure increase. That is why the problem of MH intrinsic cyclic stability becomes the most important issue in the course of development of high-pressure MH hydrogen compressors.

The most extended experimental studies of degradation effects during the cycling (up to 90,000 thermal cycles under H_2 pressure; T = 40–200 °C) were presented in work of M. Groll et al. [76] and [77]. The reversible H capacities as a function of number of cycles were determined for several AB_5- and AB_2-type alloys. Various investigations (pressure–composition isotherms, TDS, XRD, magnetisation, laser granulometry, SEM/EDS) were performed in order to determine the degradation and regeneration mechanisms involved. The reversible storage capacity of the AB_5 alloys (A = La or Mm; B = Ni, Al, Mn, Co, Sn) decayed during the cycling. This effect is stronger at higher temperatures and pressures. However, the original capacity of the materials could be recovered by heating to ~400–500 °C in vacuum leading to the decomposition of binary hydrides and recombination of intermetallides. Bowman et al. [78] observed nearly full recovery of highly degraded $LaNi_{4.78}Sn_{0.22}$ hydride following a nominal 3 h annealing at ~675 K under circa 1-bar hydrogen pressure. In contrast, the AB_2 alloys (A = Ti, Zr; B = Cr, Mn, Fe, V) showed no degradation after 42,400 cycles [76]. Iosub et al. [79] observed little or nearly isotropic broadening of the X-ray diffraction peaks of AB_2 alloys that they attributed to the reduced defect formation upon the hydrogen absorption compared to the behaviour of the AB_5 hydrides. This may account for the greater stability exhibited by hydride phases of the AB_2 intermetallics.

The prolonged H sorption–desorption cycling can impair hydrogen sorption properties even for the systems which do not undergo the disproportionation. Indeed, a 20% reduction in H sorption capacity accompanied by increase of hysteresis was observed for vanadium hydride [65] during 1000 absorption–desorption cycles performed between 24 and 135 °C. The most probable reason for that was assumed to be sintering effects accompanied by grain growth and strain relaxation. In chemically related V–Ti–Fe BCC alloys a 40% total decrease of the reversible H capacity was observed after 400 cycles at 10 bar H_2 and 20–600 °C. The cycling was accompanied by a BCC–BCT transition and by the formation of amorphous phase in the MH matrix [80].

During the last decade the material degradation issues for the MH compression materials were studied by Golben and DaCosta [81], Bowman et al. [82], [83] and [84], Laurencelle et al. [85], Li et al. [86] and [87]. It was shown that disproportionation resistance of AB_5-type intermetallides increases with increase of the binding energy between the metal atoms, and with introduction of the additives strengthening this interaction (e.g., tin-substituted $LaNi_{5-x}Sn_x$), thus resulting in the improvement of the cycle stability [88].

Structure and Morphology

Although structural and morphological features of hydride-forming alloys and hydrides are mostly considered as fundamental properties of the MH materials, some of these properties are directly related to the MH compression applications.

Fig. 5 presents the structures of typical hydrogen storage alloys following their classification shown in Table 1. As it can be seen from the last column, the hydrogenation is accompanied by a significant volume increase of the solid materials; the lattice expansion, V/V_0, typically varies from 15 to >30%. Accordingly, the cyclic hydrogen absorption and desorption is accompanied by the periodic changes in the volume of MH material loaded into a container for hydrogen compression.

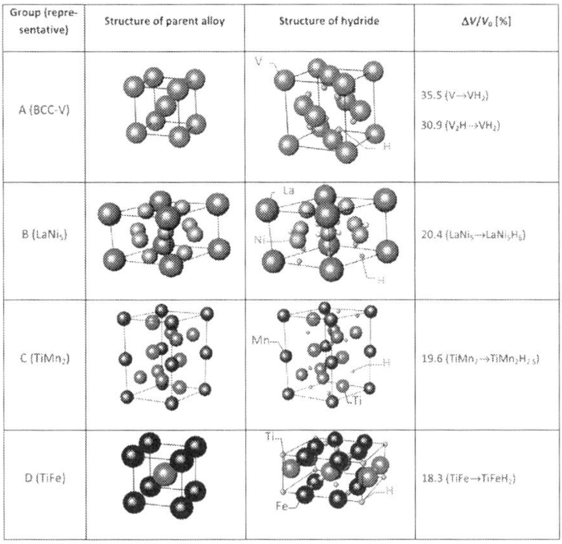

Group (representative)	Structure of parent alloy	Structure of hydride	$\Delta V/V_0$ [%]
A (BCC-V)			35.5 ($V \rightarrow VH_2$) 30.9 ($V_2H \rightarrow VH_2$)
B (LaNi₅)			20.4 ($LaNi_5 \rightarrow LaNi_5H_6$)
C (TiMn₂)			19.6 ($TiMn_2 \rightarrow TiMn_2H_{2.5}$)
D (TiFe)			18.3 ($TiFe \rightarrow TiFeH_2$)

Figure 5: Crystal structures of parent and hydrogenated alloys used for MH H_2 compression.

It is known that insufficiently high filling fraction of the powdered MH material in the MH container results in increase of the "dead space" that significantly decreases H_2 compression productivity, especially, at high discharge pressures [89]. In addition, the increase of the MH packing density is expected to result in the enhancement of the hydride effective thermal conductivity [90]. On the other hand, too high filling density, exceeding 61% of the material density in the hydrogenated state, is detrimental for the operation safety as the lattice expansion during hydrogenation can generate high stresses in the MH bed and, in turn, deform or destroy the container [91].

Fig. 6 illustrates the deformation of a stainless steel vessel following extended absorption–desorption cycling of powder LaNi$_{4.78}$Sn$_{0.22}$ hydride between ~295 K and 465 K. Thus, the filling of MH material into container for hydrogen compression is always a compromise between achieving the best operation performance and fulfilling safety requirements. The data describing lattice expansion during the hydrogenation (V/V_0) is very important for the optimisation. Since hydrides used for the H_2 compression are unstable at ambient conditions, their structural analysis requires use of *in-situ* neutron powder diffraction (NPD) and *in-situ* synchrotron X ray diffraction (SR XRD) allowing to directly monitor phase transformations during the hydrogenation–dehydrogenation. *In-situ* NPD and SR XRD were successfully applied in the detailed studies of a number of hydride systems including ones based on IMC's suitable for high-pressure hydrogen compression [92] and [93]. Conventional XRD of starting alloys and their hydrides can also be used for the determination of their structural properties, including real (crystal) densities, but its application to probe the unstable hydrides requires their stabilisation by, e.g., exposure to CO, SO_2 [52], or air at liquid nitrogen temperature [94].

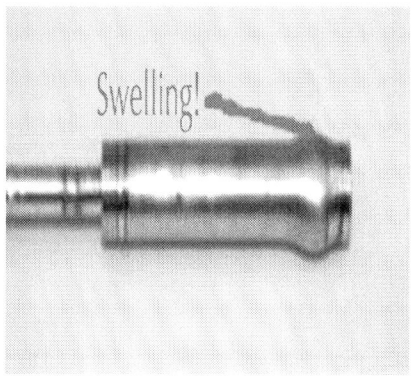

Figure 6: Swelling of a 316L stainless steel reactor vessel produced by pressure–temperature cycling of 15 g of $LaNi_{4.78}Sn_{0.22}Hx$[R.C.Bowman, Jr, previously unpublished].

Evolution of particle size and shape distribution in the course of cyclic H absorption–desorption processes is a very important factor which determines effective thermal conductivity (ETC) of the powdered MH beds. Recent findings using granular effective medium theory [90] allowed quantifying the interrelation between the morphological features of hydrogenated AB_2-type alloy and heat transfer characteristics of the corresponding MH beds. It opens perspectives for the optimisation of the MH containers for hydrogen compression towards increase of the ETC and, in turn, improvement of their dynamic performances (see also section 3.3.1).

Recently, there were published the data of the detailed experimental study of the influence of cyclic swelling of MH bed on the basis of Ti–V–Cr BCC alloy on the mechanical stresses in the containment, as well as on the changes in MH porosity and their evolution during cyclic hydrogenation/dehydrogenation[95].

APPLIED ASPECTS

General Layout

Overviews of the general layouts (Fig. 7) of the MH hydrogen compressors were presented in patent descriptions by the authors of the present review [96] and [97].

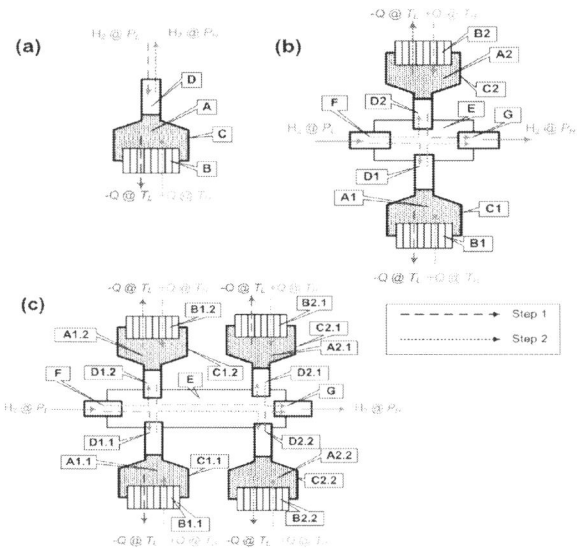

Figure 7: General layouts of MH compressor: (a) – periodically operated, (b) one-stage continuously-operated, (c) – two-stage continuously operated.

The simplest apparatus realising thermally driven hydrogen compression using MH is shown in Fig. 7(a). Metal hydride material (A) thermally coupled to a heat supply/removal accessory (B) is placed into a pressure container (C) comprising a gas pipeline (D) which allows supply or removal of hydrogen gas to/from MH (A). The gas pipeline (D) can have a built-in filter element (not shown) which provides a uniform hydrogen distribution within the MH bed, and also prevents contamination of gas pipelines with fine powder of the MH. The assembly A–D called the metal hydride compression element, or generator-sorber, provides periodic suction of low-pressure hydrogen (H_2 @ PL) when the MH is cooled ($-Q$) down to the lower temperature, TL, followed by a discharge of high-pressure hydrogen (H_2 @ PH) in the course of heating ($+Q$) of MH to the upper temperature, TH. This solution first patented in 1970 by Wiswall and Reilly[66] allows periodically operated hydrogen compression that restricts its application from the continuous technological processes.

The simplest continuously-operated metal hydride hydrogen compressor (Fig. 7(b)) comprises two compression elements (A1–D1, A2–D2) similar to the one shown in Fig. 7(a). The gas pipelines D1 and D2 are connected to a gas distributing system (E) equipped with a port (F) for the supply of hydrogen at low pressure, PL, and a port (G) for the output of hydrogen at high pressure, PH. The operation of the compressor includes two steps, 1 and 2. During Step 1 the heat supply/removal accessory (B1) of the first compression element provides heat removal ($-Q$) from the MH (A1) at a lower temperature level, TL; simultaneously, the accessory (B2) of the second compression element provides heat supply ($+Q$) to the MH (A2) at a higher temperature level, TH. During the next Step 2 the heating/cooling modes of the accessories B1 and B2 are reversed, so that B1 operates in the heat supply, and B2 in heat removal mode. Thus, a periodic reversal of the operating modes of the heat supply/removal accessories B1 and B2 synchronised with switching gas flows by the gas distributing system (E) provides the continuous operation resulting in the suction of low-pressure hydrogen to the port F and the release of high-pressure hydrogen from the port G.

An approach to generate high H_2 pressures at modest operating temperatures is the use of multi-stage hydride compressors, a concept developed at Ergenics Inc. [98]. The multistage compressor uses a series of two or more alloys differing by thermal stabilities of their hydrides. Fig. 7(c) shows an example of layout of two-stage MH compressor. The alloy forming the most stable hydride is placed in the compression elements of the first stage (A1.1, A1.2), and other MH are loaded to the compression elements belonging to the next stages, in the order of decrease of their thermal stability (A2.1, A2.2). The multistage operation allows achievement of higher overall compression ratios using the same or smaller temperature swing. For example, five-stage MH compressor developed by Ergenics allows H_2 compression from 7 to 250 bar in the temperature range 30–90 °C with water as a heating/cooling agent [99].

The gas distributing system (E) can be made as a one-way (check) valve arrangement (see Fig. 8 as an example); the periodic heating/cooling of heat supply/removal accessories (B) is conveniently controlled by timing relays [100], [101] and [102].

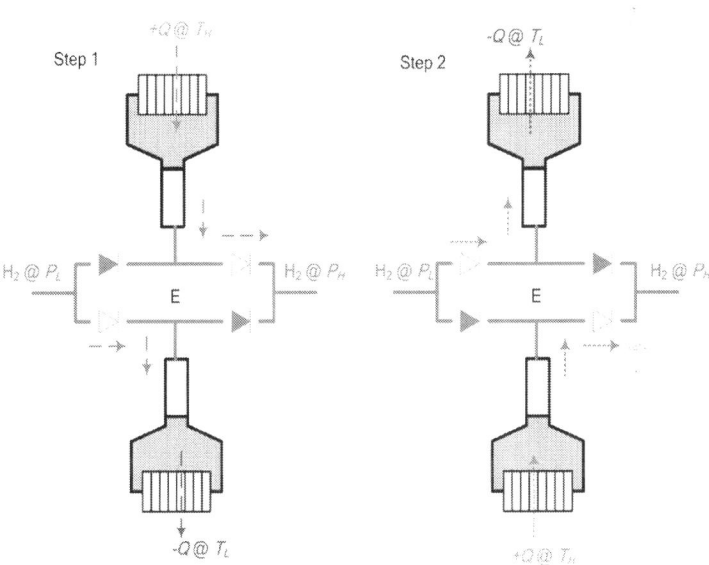

Figure 8: Operation of one-stage continuously operated MH compressor (Fig. 7(b)) when gas-distributing system (E) is made as a check valve arrangement. Opened and closed check valves are shown as empty and filled symbols, respectively.

The basic engineering approach described above is presented in a number of publications and patents. Before considering the details of its implementation (section 3.3), we would like to present thermodynamic analysis of the MH compressors as heat engines (next section), and to discuss efficiency of the compression.

MH H$_2$ Compressors as Heat Engines

A detailed thermodynamic analysis of an MH hydrogen compressor (MHHC), or MH thermal sorption compressor (MH TSC), as a heat engine has been performed by Solovey [5], [103] and [104]. Influence of various factors on thermodynamic performances of the MH TSC's was also considered in Refs. [57], [89],[105], [106], [107], [108], [109] and [110].

Hydrogen compression in an ideal MHHC/MH TSC (Fig. 9) is achieved by sequential processes which include:

- isobaric–isothermal absorption of low-pressure hydrogen ($P_1 = PL$) at a lower temperature, TL (1–2);
- polytropic heating of the MH from lower (TL) to higher (TH) temperature (2–3);
- isobaric–isothermal desorption of high-pressure hydrogen ($P_2 = PH$) at higher temperature, TH (3–4);
- polytropic cooling of the MH from TH to TL and isobaric cooling of high-pressure hydrogen from TH toTL (4–5–1).

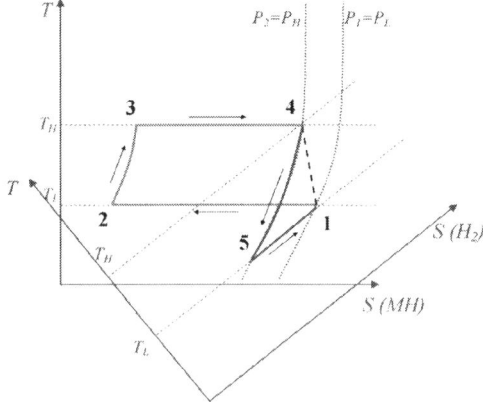

Figure 9: Entropy (for both metal hydride and hydrogen gas)–temperature diagram of the operation of an idealised MH compressor [104].

When pressure–temperature dependence for H_2 absorption/desorption is described by the van't HoffEquation (2), the heat, Q, will be transformed to compression work, W, with the efficiency of Carnot cycle (C) realised within the same temperature range:

$$W = Q \frac{T_H - T_L}{T_H};$$

(8a)

$$\eta_C = \frac{W}{Q} = \frac{T_H - T_L}{T_H};$$

(8b)

where T_H and T_L are the temperatures of heat supply and heat sink, respectively.

Equation (8b) gives the upper limit of the MH TSC efficiency. The efficiency of the real engine will be reduced by two groups of factors described below.

- The first group includes intrinsic factors characterising physical–chemical behaviour of the real system of hydrogen gas with hydride-forming materials. These are reversible hydrogen sorption capacity, plateau slope, hysteresis, and dilatation.

- The second group is related to the design and technological performance of the specific MH TSC's and their components (mainly, MH containers). These include rates of heat exchange between the heat transfer fluid and the MH bed, volume of the dead space (void fraction in the MH bed), material consumption of the MH container (containment/MH material weight ratio), and efficiency of heat recovery.

Reversible hydrogen storage capacity, C, is a difference between hydrogen concentration in the MH at suction, $C_L(T_L, P_L')$, and discharge, $C_H(T_H, P_H')$, conditions (Fig. 3). It is determined by direct measurements of the pressure–composition isotherms for the MH alloy at TL and TH. If the detailed experimental PCT data for the MH are available, they can be fitted using a model for the PCT diagrams, and then the values of C for different operating conditions can be easily evaluated (see Section 2.1). This factor influences the MH TSC efficiency both directly and, furthermore, via such engineering factors as dead space and material consumption. Increasing reversibility of the hydrogen storage capacity leads to the smaller differences between the MH TSC efficiency and its ideal (Carnot) value. These two values converge for a completely reversible H absorption–desorption. The thermodynamic properties of the MH–hydrogen system are important both for adjusting hydrogen suction/discharge pressure to the available temperature range and for the achieving of the best possible efficiency of the MH TSC. First of all, this concerns the specific heat $Q \approx \mid H^0 \mid$ of hydrogen sorption/desorption which determines thermal energy consumption during hydrogen compression. Specific heat capacity, cp, of the MH alloy is also very important.

It should be noted that the correct calculations of the high-pressure MH TSC parameters should also take into account fugacity of hydrogen and temperature variation of the differences in heat capacities of the reagents participating in the Reaction (1). Neglecting these important features can introduce large errors when using the Van't Hoff Equation (2) values for the discharge pressures; these deviations can be up to 30% higher than the correct ones [89].

As briefly mentioned in the Section 2.1, in the plateau region the equilibrium pressure of hydrogen absorption exceeds the equilibrium pressure of hydrogen desorption. This factor, sorption hysteresis, is expressed quantitatively by a difference in free energy [13]:

$$\Delta G_{hyst} = RT \ln\left(\frac{P_A}{P_D}\right) \quad \text{or}$$

$$\delta_T = \ln\left(\frac{P_A}{P_D}\right) = \frac{\Delta G_{hyst}}{RT} \qquad ; \tag{9}$$

where P_D and P_A are, respectively, desorption and adsorption hydrogen equilibrium pressures measured at the same temperature, and δT is isothermal hysteresis factor. This value is the characteristic of a specific MH–hydrogen system and should be determined experimentally. As it was shown in Ref. [103], hysteresis causes additional energy consumption required to close the thermodynamic cycle of an MH TSC. It causes losses in hydrogen compression work and, therefore, reduces the efficiency of hydrogen compression. In thermodynamic calculations of the MH TSC efficiency the hysteresis can be described by isobaric factor, δP which can be expressed as:

$$\delta_P = \frac{1}{T_A} - \frac{1}{T_D} \approx \frac{\Delta S_D^0 - \Delta S_A^0}{Q_S}; \tag{10}$$

where index A corresponds to absorption, index D for desorption and $Q_s \approx -\Delta H^0{}_A \approx -\Delta H^0{}_D$.

Another intrinsic factor indirectly influencing on the efficiency of the MH TSC is dilatation that is relative volume change resulting from expansion of the parent metal lattice in the course of the MH formation (seeSection 2.3.4). The typical values of the dilatation coefficient, $\alpha = \Delta V / V_0$, are of 10–30% [6], see alsoFig. 5. Dilatation causes changes in the MH density that influences on the value of the dead space in the MH TSC (see below) and, on the other hand, results in swelling of the MH bed that affects safety and reliability of MH containers.

The rate of heat transfer between the heat carrier and the MH bed is the most important design and technological factor affecting both the efficiency and productivity of the MH TSC. In efficiency calculations this factor is taken into account by taking TH below the heating temperature and TL above the cooling one; the differences are calculated starting from the effective overall heat transfer coefficients [105].

The disperse structure of the powdered MH bed, as well as presence of voids filled by gas in the MH container and gas distribution system determine the effect of the dead space, which negatively influences the MH TSC efficiency and productivity [45]. This negative influence increases with increasing discharge pressure. For example, at output pressure of 300 bar and typical value of a dead space of 0.25 cm^3 per 1 g of the MH alloy having hydrogen sorption capacity ~140 cm^3/g STP, the fall in both productivity and efficiency reaches 30% [89].

Influence of the dead space on the efficiency of MH TSC can be taken into account by introducing dimensionless coefficient KV:

$$K_V = \frac{m_H(o) - m_H(i)}{m_H(MH)};$$

(11)

where $m_H(o)$ and $m_H(i)$ are the weight (or number of moles) of H$_2$ in the dead space at output (discharge) and input (suction) conditions, respectively, and $mH(MH)$ is the weight (or number of moles) of H$_2$ in the metal hydride (equal to the reversible hydrogen sorption capacity multiplied by the weight of the MH). K_V depends on conditions of hydrogen suction and discharge and, also, on the real and packing densities of the MH. Decreasing the dead space can be achieved first of all by increase of the MH bed packing density, taking into account safety requirements originated from swelling.

An important design factor affecting the MH TSC efficiency is material consumption of the MH container. It includes pressure containment and heat exchanger; these, being periodically heated/cooled, cause heat losses and decrease the MH TSC efficiency. The material consumption can be taken into account by introducing coefficient K_M equal to the ratio of the total weight of the empty container to the weight of the MH therein. The negative influence of the material consumption can be partially mitigated by the heat recovery.

The quantification of the influence of the above-mentioned factors on the efficiency of the MH TSC, η, was derived by Solovey [5] and [103] as:

$$\eta = \frac{Q(\eta_C - \delta_P T_L)(1 - K_V)}{Q(1 - K_V) + (1 - \sigma)\left[c_\Sigma \frac{1+K_M}{\Delta C}(T_H - T_L) + QK_V\right]} \tag{12}$$

Here the numerator represents the net heat required for hydrogen compression, and denominator is the total supplied heat. Q is the net heat required for hydrogen desorption from the M_H; ηC is the Carnot efficiency (Equation (8b)); δ_p is isobaric hysteresis factor (Equation (10)); K_V is the dead space coefficient (Equation (11)); σ is heat recovery efficiency; cΣ is the total heat capacity of the MH container with MH bed; K_M is material consumption coefficient, and ΔC is reversible hydrogen sorption capacity expressed as hydrogen weight fraction in the M_H.

Efficiencies of the MH TSC calculated using Equation (12) at various heat recovery efficiencies are shown in Fig. 10. The figure also contains our estimations of the efficiency range of industrial mechanical compressors produced by RIX Industries [111]. These compressors having the productivity of 50–100 m³/h and compression ratio of 50–350 are characterised by the efficiency of 40–45%. This is superior to the efficiency of the MH compressors (below 25% at TH ~ 150 °C). However, mechanical compressors also require significant investments from their operators and require much more directly generated electrical energy than the concept of the MH TSC that is based on utilisation of the waste thermal energy.

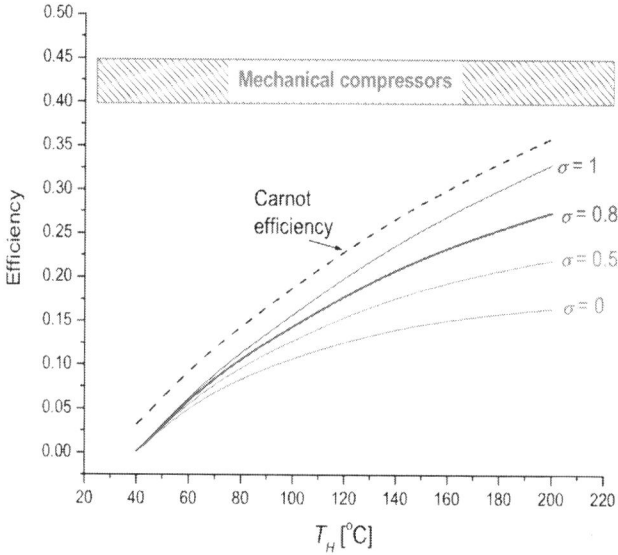

Figure 10: Efficiencies of MH TSC using LaNi$_5$ at TL = 30 °C and different heat recovery efficiencies, σ: ΔC = 1.4 wt.%; Q = 31 kJ/mol H$_2$, or 217 kJ/kg (MH); δP = 2·10^{-4} K^{-1}; KV = 0.1; KM = 1.0; c_f = 0.51 kJ/K.

Calculations presented above assume that the energy inputs for the MH H$_2$ compression are associated only with the heating of the MH material from TL to TH, and the cooling from T_H to T_L is a spontaneous process of the heat dissipation into environment. If the cooling requires additional energy input (e.g., when a heat pump is used), the amount of consumed energy will be higher resulting in a decreased efficiency. Calculations by Kelly and Girdwood [110] for the H$_2$ compression from P_L = 130 bar (T_L = 30 °C) to P_H = 414 bar (T_H = 130 °C) yielded the efficiency of the process (related to the isothermal compression work) as 2.9%, or 11.6% of the Carnot efficiency. According to our calculations using Equation (12) (no heat recovery) and parameters presented in Ref. [110], the corresponding values are 7 and 28%, respectively. The origin of the difference is in the accounting of the energy input used for the cooling (about 51% of the total energy consumption) applied in Ref. [110].

Low energy efficiency is a common feature of heat engines operating in a narrow temperature window ((8a) and (8b)). Various energy losses in the real MH compressors result in further decrease of their efficiency. Finally, as it was shown in the previous paragraph, the efficiency further decreases because of additional energy inputs. Therefore, the MH hydrogen compression can become beneficial either for some special applications, or when the energy inputs are associated only with low-grade waste heat.

Analysis of the value/exergy of primary energy inputs consumed in the thermally driven MH H_2compression can be a useful tool in the comparison of this method with conventional compression technologies.

Fig. 11 shows calculated exergy efficiencies of the MH and mechanical electrically driven hydrogen compressors [106]. There the single-stage MH TSC based on $LaNi_5$ intermetallic compound, and two-stage MH TSC using $LaNi_5$ for the first stage and $Ce_{0.5}La_{0.5}Ni_5$ for the second one were considered. The total heat capacity of the containment was assumed to be equal to the heat capacity of the MH ($KM = 1$), and the efficiency of heat recovery (σ) was assumed to be 0.5. It can be seen from Fig. 11 that MH hydrogen compression can provide the efficiency gain over mechanical one at the temperatures below 200 °C for a single-stage MH TSC and below 100 °C for a two-stage MH TSC. For these conditions a significant energy benefit in comparison with mechanical compression takes place. It also can be seen that an increase in the number of stages of an MH compressor results in a significant decrease of its efficiency. Thus, the number of compression stages for large-scale industrial applications should be minimised.

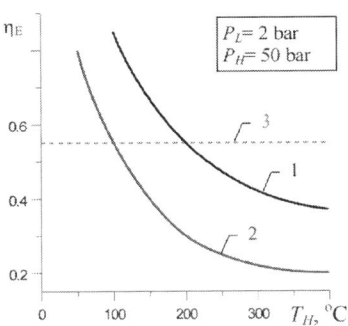

Figure 11: Exergy efficiency of single- (1) and two-stage (2) MH compressor and mechanical H_2 compressor (3) as function of heat source temperature [106].

The contribution of factors influencing the exergy efficiency of the MH compressors is shown inFig. 12[104]. The efficiency of heat transformation into the energy of the compressed H_2 was found to be about 0.58 that corresponds well to the data for one stage MH compressor at $TH \sim 100\ °C$ (Fig. 11). It was also noted in Ref. [104] that if to consider exergy efficiency of electrically-driven mechanical compressors starting from the heating value of a fuel burnt at a thermal power plant (efficiency about 0.4) then the efficiency will be about 0.18, or 3.2 times lower.

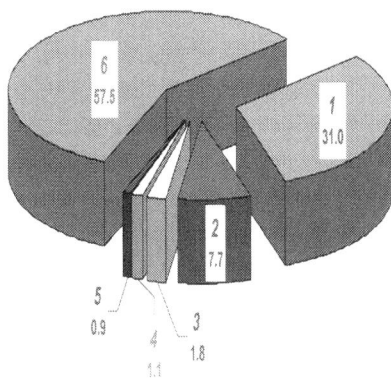

Figure 12: Exergy balance (in % to the exergy of heat at TH) for H_2 compression using MH [104]: 1 – losses in transient process; 2 – losses caused by heat transfer; 3 – losses for dead space; 4 – losses due to heat transfer with the environment; 5 – losses in gas distribution system; 6 – useful exergy.

As can be seen from Fig. 12, the major factors contributing to the decrease of the efficiency of MH compressors are transient losses (1) and losses caused by heat transfer (2). Reduction of these losses is the main objective of further improvements in design and operation features of the MH hydrogen compressors which will be considered in the next section.

Design Features and Performances

Improvement of the basic engineering approach in the development of MH hydrogen compressors (section 3.1) mainly concerns:

- • (Optimisation of the hydride-forming alloys;
- • (Design of compression elements;
- • Methods and accessories for heat supply/removal;
- • Number of compression elements, their gas connections and the sequence of operation of the associated heat supply/removal accessories.

The approach to the selection/engineering of the MH materials, mainly related to the adjustment of the required operating pressures over the available temperature range, was considered in section 2. The present section will consider the engineering solutions of MH containers to be critical components of the MH compressors (section 3.3.1) and system features followed by a brief overview of state of the art in the international development of MH compressors (3.3.2).

MH Containers/Compression Elements

A proper design of the containment for the MH material for H_2 compression, together with the associated H_2 gas and heat supply/removal accessories, has two main objectives. First of all, it aims at the achievement of high hydrogen charge–discharge rates to provide shorter cycle time and higher productivity of the compressor. Secondly, it has to provide higher efficiency of hydrogen compressors by reducing the losses (see Fig. 12).

The main problem to be solved for the achievement of both goals is the intensification of heat transfer between the heat supply/removal accessories and the MH material. The main factor limiting the H_2 charge/discharge dynamics of the MH containers is the low thermal conductivity of the powdered hydride beds [45]. Moreover, its value is strongly related to both design and technological parameters (geometry, MH packing density, as well as the wall heat transfer resistance), and on the operation conditions. Usually, the effective thermal conductivity of a powdered MH bed can vary in the range of 0.13–2.3 W/(m K) while the thermal conductivity of a bulk alloy is more than order of magnitude higher, e.g. 30 W/(m K) for $LaNi_5$[7].

Optimisation of the MH bed heat transfer performances requires their modelling and verification by comparison with experimental data. The MH bed heat transfer modelling (with a subtask of hydrogen mass transfer) was developed rather intensively during the last three decades.

Various computation and experimental approaches are presented, for example, in Refs. [47], [50], [90], [112], [113], [114], [115],[116], [117], [118], [119], [120], [121], [122], [123], [124], [125] and [126].

A conventional way to improve the heat transfer characteristics of the MH bed is in increase of the surface area of heat exchange and reduction of the characteristic heat exchange distances. It can be done, for example, by using long tubular MH containers of a small diameter used in the compressor where simultaneously heated/cooled containers are immersed into one heating/cooling jacket [101]. The "shell-and-tube" solution can be used for both separation/purification and compression of hydrogen [127]. Application of tubular containers, 12–25 mm in outer diameter filled with 150–900 g of MH powder allows to achieve a reasonable duration of H_2 absorption/desorption (half-cycle time), 5–10 min [58], [75] and [128]. The necessary hydrogen storage capacity can be achieved by connecting several containers in parallel; as example, compression element of industrial scale MH compressor (up to 14 kg of MH) comprises of sixteen tubular MH containers immersed into a common heating/cooling jacket [58]. Additional intensification of the heat exchange between the heating/cooling fluid and the external surface of the MH containers is achieved by use of fins [75], or thermally conductive metal blocks [128].

Effect of the aspect ratio of the tubular MH containers on their performances was studied in Ref. [124]. The issues of modelling and optimisation of multi-tubular MH beds were considered in Ref. [117]. A typical engineering solution of hydrogen storage and supply device composed of tubes filled with MH is presented in a patent [131].

Apart from a decrease of characteristic heat transfer distance, the decrease of the diameter of the tubular MH container allows to reduce the material consumption because a pressure vessel with a smaller diameter can have a thinner wall for the same pressure rating. This results in a significant increase of the efficiency, due to reduction of transient losses for periodic heating/cooling of the containment (see section 3.2).

The best realisation of the approach described above was achieved by Ergenics [129], [130], [132], [133],[134], [135] and [136]. Metal hydride material is loaded into a tubular containment having a small outer diameter, down to 1/16", or 1.588 mm. The intensification of heat exchange between the outer surface of the hydride tube and heating/

cooling fluid can be achieved by fins formed by steel wire wound and soldered onto the hydride tube [132], [133] and [134]. It allows hydride beds to be thermally cycled at a rapid rate (<1 min) resulting in high productivities. The outer "spring" formed by the wound wire also reinforces the hydride/hydrogen containment allowing for its safe operation at high pressures. Alternative solutions[129] and [130] include placement of the "spring" inside the cylindrical containment, so the MH becomes located in between the inner surface of the cylinder and the outer surface of the spring; thus allowing compensation of the MH swelling effects during the hydrogenation.

A plurality of the hydride tubes can be assembled in an MH reactor combining features of the hydrogen manifold and heat exchanger (Fig. 13). The designs of the reactors are modular, resulting in a high volume low cost production.

Figure 13: 1/16″ OD hydride tube ring manifolds (a), the manifolds stacked into hydride heat exchanger (b), the placement of the heat exchanger in MH compression element (c), and two-element 3-stage MH compressor assembly (d). Adopted from Linde–MRT–Ergenics joint presentation [135].

Intensification of heat transfer in the larger MH containers can be achieved by a placement of the MH material within a heat transfer matrix inside the containment. The MH containers have built-in heating and cooling accessories thermally coupled with the heat transfer matrix, as well as the pipelines for supplying hydrogen gas to and receiving hydrogen gas from the MH. These elements are present in numerous developments of the MH containers.

A typical approach is shown in Fig. 14. It is used for the medium-scale MH-based hydrogen storage and compression. The MH container is made as a cylindrical gas-proof containment equipped with end caps. The MH material is placed inside the containment (1) into a heat transfer matrix to form a metal hydride bed (2). Hydrogen input/output (3) is provided by an axial pipe (3.1) installed at one end cap and usually ended by an inline gas filter (3.2). Heating and cooling is provided by a heat transfer fluid (e.g., water) running through either external heating/cooling jacket (A) or core tube of the inner heat exchanger (B) (4).

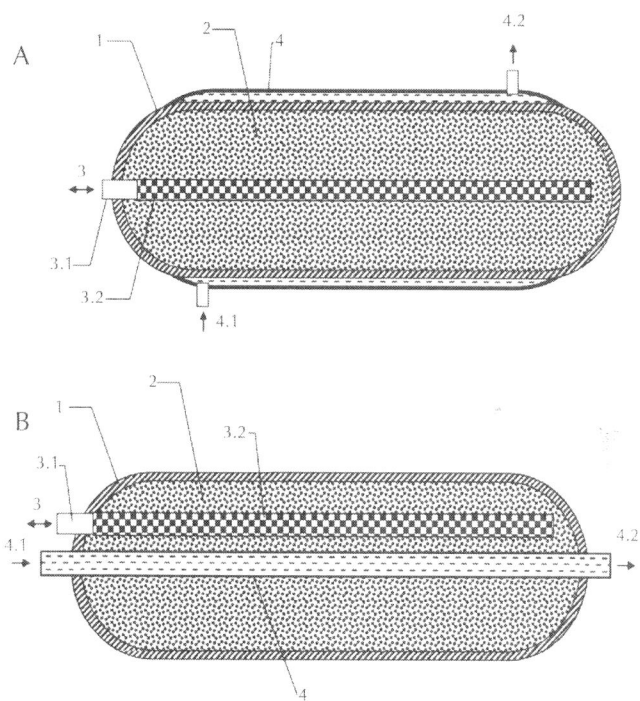

Figure 14: Typical layouts of the MH containers with external (A) and internal (B) heating/cooling by a flow of heat transfer fluid: 1 – gas-proof containment; 2 – metal hydride bed (MH material + heat transfer matrix); 3 – H_2 in/out line: 3.1 – H_2 inlet/outlet pipeline, 3.2 – gas filter; 4 – heating/cooling jacket (A) or core tube of inner heat exchanger (B): 4.1 – input and 4.2 – output of heat transfer fluid.

The main solutions utilising such an approach include:

- Type of the heating/cooling:

  flow of heat transfer fluid for both heating and cooling [97], [98], [99], [100], [108], [109], [110],[137] and [138]; this solution is used in most cases [58], [75], [101], [129], [130], [131], [132],[133] and [134];

  electric heating, in combination with convective (natural or forced) air cooling [66], [139],[140] and [141];

  electric heating, liquid cooling [71], [102], [128], [139] and [140];

  electric heating, cooling using a cold radiator thermally coupled with MH bed through a gas-gap thermal switch [142] and [143];

  heat-pumping systems including thermoelectric (Peltier) modules [96] and [144];

  heat pipes, in combination with electric heaters or catalytic combustors (heating) and flow of liquid or gaseous heat transfer fluid (cooling) [145] and [146].

- Placement of the heating/cooling means with respect to the containment:

  external [58], [75], [96], [109], [117], [118], [137] and [140];

  internal [94], [95], [96], [97], [125], [126], [145] and [146];

  combined [102], [128], [140], [141] and [142].

- Type and layout of the heat transfer matrix, including:

  heat-conductive fins [10], [94], [97], [125], [126], [137], [145], [146], [147], [148], [149], [150],[151] and [152];

  coiled tube heat exchanger [126], [147] and [153];

  metal foams [10], [124], [138], [142], [154] and [155] or honeycomb metallic structures [122]; an alternative arrangement of the metal foams with the fins was shown to be very efficient [10];

a simple and efficient method of the forming of the MH bed is in the compacting of the powders of an MH and a heat-conductive material, including porous metals [109], [156], [157] and [158] or expanded natural graphite, ENG [159], [160], [161] and [162]. An alternative arrangement of the MH/ENG compacts with the fins was patented in Ref. [163]. Significant improvement in effective thermal conductivity in MH–carbon composites was recently observed in a course of direct deposition of single-wall carbon nanotubes on the surface of particles of the hydride-forming alloy [164]. The details on the preparation of the MH/ENG compacts on the basis of AB_2-type alloys suitable for high-pressure applications were recently reported in Ref. [165].

It has to be noted that performances of the metal hydride compressors and its main element, metal hydride container, are strongly dependent on numerous factors that include the following conflicting trends[97] and [137]:

- The reduction in the size (diameter) of the container, results in better hydrogen absorption/desorption dynamic performances, and, in turn, in the shortening the operation cycle time. Correspondingly, the specific productivity per unit of weight of the MH material is increased. At the same time, due to kinetic limitations (see section 2.2), this improvement has a maximum limit, and it seems unfeasible to achieve the half-cycle time shorter than 1–2 min. Thus, high total output productivity for smaller containers can be achieved by the increase of their number in a compression element. However, large increase in the number of the hydride containers will result in the increase of the number of joints in the corresponding gas manifolds and will also increase the probability of leaks that has a drawback from the safety and reliability viewpoints.

- Although shortening of the cycle time increases the productivity of the MH compressor, it has a negative influence on the operation lifetime, due to the degradation of the MH (section 2.3.3). The service life of an MH material at specified pressure/temperature conditions is determined by a number of hydrogen absorption/ desorption cycles ($\sim 10^4$ when operated in pure hydrogen at T < 200 °C and PH_2 < 200 bar), and for the shorter times of the operation cycle the lifetime of the compressor will be decreased.

- The required total output productivity can be also increased by the increase of the amount of MH material in the compression element, particularly, by the increase of the size of the MH container. However, the drawback is a longer operation cycle. Introduction of special heat distribution means/heat transfer matrix results in an increase of the material consumption and, in turn, the total heat capacity of the MH container. The same effect has the increase in the diameter of MH container itself, since to withstand the operating pressure the containment should have thicker walls that results in the increase of the total heat capacity. It significantly increases transient losses and reduces the efficiency (section 3.2). Reducing the weight of the high-pressure containment can be achieved by the application of a "hybrid" solution where MH material is distributed inside a composite cylinder having a light-weight multilayer structure. Usually, this solution is used for weight-efficient hydrogen storage[166] and [167], but modifications that can be adopted for MH hydrogen compression are also known[137], [145] and [146]. Mainly the adaptations are related to the reduction of the dead space in the inner volume of the MH container.

• Decrease of the dead space is very critical for increasing efficiency of high-pressure hydrogen compression applications [89]. First of all, it is achieved by the increase of the filling fraction for the MH material. However, due to the "swelling" effects, there exists an upper limit (61% of the density of the hydrogenated material) of the filling density to provide safe operation (see section 2.3.4). The secondary effect is in further pulverisation of MH in the course of hydrogen absorption/desorption cycling that causes concentration and agglomeration of MH particles in the lower parts of the containment [168]. The common way to mitigate this effect is to keep rather large length/diameter ratio and to place the container horizontally.

In summary, we can conclude that any particular realisation of MH containers/compression elements strongly depends on the specific application requirements.

For the smaller-size applications, operation performances (first of all, productivity) are most important, and the corresponding solutions envisage forced heating and cooling of thin ($\leq$10 mm) MH beds incorporating MH material and heat distribution means (metal foams/fins). Fig. 15 presents an example of the MH container developed by Jet Propulsion Laboratory/NASA as a compression element for 0.6–50 bar, 260 L H_2/h hydrogen compressor for hydrogen sorption cryocooler used in on-board hydrogen Joule–Thomson cryocoolers for the ESA Planck mission [142]. The compression element provides fast desorption of the high pressure H_2 due to electric heating of the

MH bed (615 g of LaNi$_{4.78}$Sn$_{0.22}$ dispersed in aluminium foam). The cooling is provided by a cold (<0 °C) radiator thermally coupled with the outside of the container via 0.75 mm thick gas gap heat switch. The latter couples or isolates the bed with the radiator, by the variation of the pressure of H$_2$ gas (~10^3–10^{-2} Pa) using periodic H$_2$ desorption–absorption by ZrNi intermetallic alloy. Further intensification of the MH cooling can be achieved by, for example, use of thermoelectric coolers/Peltier modules [96].

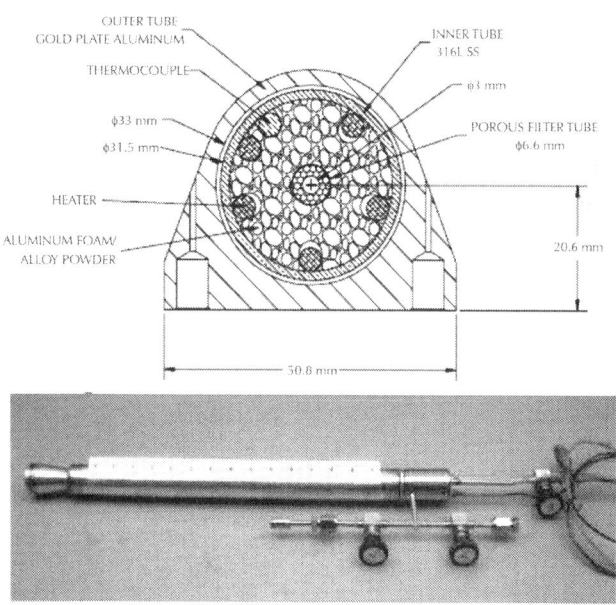

Figure 15: Cross section (top) and general view of MH container/compression element for space applications [142].

For the medium- and large-scale applications efficiency and manufacturing cost become crucial. It poses a motivation for the usage of the available waste heat sources such as steam and hot water. The corresponding solutions, as a rule, include quite large MH containers (≥10 kg MH) heated/cooled by a flow of heat transfer fluid (hot/cold water or oil, steam/water, etc.). Typically, heat distribution is provided by heat conductive fins disposed within the MH powder [94], [97], [137], [145] and [146]. Fig. 16 presents an example of the MH container for H$_2$ compressor (up to 200 bar) that comprises of 12–15 kg of MH powder (1500 to 2000 L H_2 STP storage capacity) and uses wet steam (up to 140 °C) or super-heated water (up to 180 °C) for the heating [94] and [97] providing up to 200 bar H_2 output pressure and hour productivity up to 1000 L H_2 STP. The container showed satisfactory hydrogen compression performances when typical half-cycle duration (H_2 absorption or desorption) was about 30 min [97]. Further increase of the size of MH container for H_2 compression (about 1 m long, 200 mm internal diameter, MH load 45 kg) was shown to result in significantly longer (>1 h) H_2 desorption time because of the less efficient H_2 mass transfer inside the container [169].

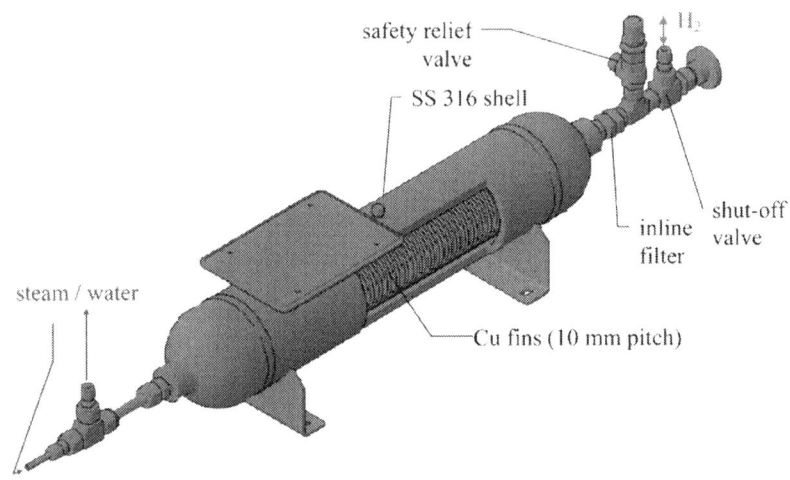

Figure 16: Metal hydride container for medium-to-large scale H_2 compression applications.

Integrated Compression Systems

Table 2 presents summary of design features and performances of the MH compressors developed since the publication of the first patent [66] describing application of MH for H_2 compression.

Table 2: MH H$_2$ compressors developed in 1970–2013

Year	Design features				Performances							Developer; notes	Ref.
	Operation	# of Stages	# of containers per stage	Hydride-forming material	TII [°C]	PI [bar]	THI [°C]	PII [bar]	Productivity [m³/h STP]	Half-cycle duration [min]	Efficiency [%]		
1	2	3	4	5	6	7	8	9	10	11	12	13	14
1970	Periodic	1	1	Hfc (60 g)	20	15	137 (200)	255 (640)	No data			US Atomic Energy Commission; electric heating, convective cooling	[66]
1971	Periodic	1	1	VHx (100 g)	18	7	50	24	0.072	1	No data	Brookhaven NL (US); water heating/cooling;	[62]
1979	Continuous	1	2	LaNi₅ (700 g)	20	25	80	20	No data	2	7.7	National Chemical Laboratory for Industry (JP); water heating/cooling, 28 W mechanical power output	[170]
1980	Periodic	1	1	LaNi₄.₆Al₀.₄ (1.5 kg)	23	12	68	2.1-3.3	No data	3-16	1.6-2.4	Sandia NL (US); water heating, cooling, water pump/151 per cycle	[171]
1983	Periodic	1	19	LaNi₅ (19 × 0.91 kg)	27	3	90?	18	21.6	1	14.2?	Tsukuba Research Centre (JP); used for desalination by reverse osmosis, water heating/cooling (30 l/min)	[148]
1996	Continuous	4-6	No data	AB₅	25	1-4	85	40-200	Up to 2.5	2	No data	Ergenics Inc (US); commercial series, 20 years life time	[172]
1993	Continuous	1	3	Hfc + 2 wt.% Mm (3 × 1 kg)	20	10	250	100	0.42	15	4-7	Universidade Estadual de Campinas (BR); water cooling, water/electric heating	[71]
1995	Periodic	1	3	LaNi₅ (3 × 1.5 kg)	25	10	570	150	14	45	39?	Inst Probl Mech Eng (US); electric heating, 1 kW in total, convective cooling, 106 dm³ volume, 16 kg weight	[139]

Year	Mode			Material								Notes	Ref.
1995	Periodic	1	7	LaNi$_5$ (7 × 1.4 kg)	25	10	370	300	0.7	120	0.92*	Inst. Probl. Mech. Eng. (UA); electric heating (8 kW in total), convective cooling; 170 dm³ volume; 142 kg weight	[139]
1996	Periodic	3	1	1 ZrNi (0.225 kg) 2 LaNi$_{4.8}$Sn$_{0.2}$ (0.92 kg) 3 LaNi$_{4.8}$Sn$_{0.2}$ (1.5 kg)	25	0.001	1 280 2 95 3 240	1 1 2 3 3 103	N/A			NASA/JPL (US); electric heating (245 W), radiator cooling hydride beds mass 57 kg; Space Shuttle flight	[173],[174] and [175]
1998	Continuous	2	30	1 LaNi$_4$Mn$_{0.5}$ (30 × 1.33 kg) 2 LaNi$_5$ (30 × 1.33 kg)	25	3	250	150	10	35	4.48*	Inst. Probl. Mech. Eng. (UA); electric heating, forced air cooling (27 kW in total); size 2350 × 1150 × 1050 mm; 700 kg weight	[141]
1999	Periodic	1	1	MmNi$_5$ (1.6 kg);	15	25	327	400^d	0.24	60	2.44*	Inst. Probl. Mech. Eng. (UA); electric heating (1 kW); air cooling	[176]
1999	Periodic	2 3	1 1	1 Hydralloy C2 (AB$_2$) 2 Hydralloy C0 (LiMn$_{1.5}$V$_{0.05}$Fe$_{0.05}$) 3 TiCrMn$_{0.3}$Fe$_{0.3}$V$_{0.1}$ 1 kg each	20 20	12 18 30	60 60	85 110 200	No data	40 60	No data	Helsinki Univ. of Technology (FI); water heating/ cooling; combined compressor and heat pump, medium temperature −30 °C heat upgrade/thermal efficiency 1.3 1.5.	[177]
2000s	Continuous	No data			25	0.8	100	300	1	No data		Industrial Technology Research Inst. (TW); Commercial series	[178]
2001	Periodic	1	1	V1x	20	100	527	5000	No data			Russian Federal Nuclear Center; air cooling/electric heating; research facility for H$_2$ and D$_2$ compression	[179]
2001	Continuous	1	No data	AB$_5$	25	20	400	345	0.33	16	No data	Ergenics Inc (US); air cooling/electric heating; prototype; dimensions of heat exchanger in MH bed D250 × 500 mm	[133]

Year	Operation	No.	Material									Comments	Ref.
2002	Continuous	6	$LaNi_{4.25}Sn_{0.55}$ (6 × 615 g)	7	0.6	197	50	0.26	60	8.68		*NASA JPL (US); chiller plate cooling/electric heating (410 W); used for the cryo cooling on board of Planck spacecraft; MH bed ~500 × 51 × 51 mm; lifetime ~20,000 cycles*	[142]
2004	Continuous	No data			1	No data	350	No data		15		*Ergenics Inc (US); heating by natural gas*	[134]
2005	Periodic	1	$MnNi_{4.6}Al_{0.4}$ (0.4 kg)	20	5	95	43.8	0.34	4.2	7.3		*Indian Inst. of Technology; liquid heating/cooling (1.3 L/min)*	[151]
2006	Periodic	1	1 $LaNi_{4.8}Sn_{0.2}$; 2 $LmNi_{4.6}Sn_{0.4}$; 3 $MmNi_{4.6}Al_{0.4}$ (25 g each)	20	1	80	20	0.02	8-20	5		*Institut de Recherche sur l'Hydrogène (CA); water heating/cooling; pre-compression of H_2 from alkaline electrolyser*	[148] and [160]
2007	Periodic	1	$Mm_{0.8}Ca_{0.2}La_{0.1}(Ni_{4.6}Al_{0.4})_2$; 1000 l H_2 capacity	25	40	170	450	2.4	15	No data		*Zhejiang University (CN); oil heating/cooling*	[181]
	Periodic	2	1 $Mm_{0.7}La_{0.3}Ca_{0.6}Ni_5$; 2 $Ti_{0.7}Cr_{1.2}Mo_{0.4}V_{0.1}$	25	40	99	450	1.2	15			*Zhejiang University (CN); water heating/cooling*	
2008	Continuous	3		20 (?)	0.5	90 (?)	100/435	15	50 s	69		*Ergenics Inc (US); liquid heating/cooling module productivity >215 l/m²*	[135] and [136]
2009	Continuous	4	1 $LaCoNi_5$ (160 g); 2 $(Ti,Zr)(Fe,Mn,Cr,Ni)_2$ (120 g)	15	7	110	200	0.06	10	1.6		*Univ. Western Cape (ZA); electric heating (400 W), water cooling*	[128]
2009	Periodic	1	1 $LaNi_{4.7}Al_{0.3}$ (3.5 kg); 2 $LaNi_{4.8}Sn_{0.2}$ (45 kg)	50	16...0.2	175	33	1.02	30/60s 190(D)s	2.3		*Tech. Univ. of Lodz (PL); oil heating/cooling; 1/1.5 model necessary for operation of H_2 hardening furnace*	[182]

Year	Mode			Material								Institution	Ref.
2009	Periodic	3	1	1 LaNi$_{4.6}$Al$_{0.4}$; 2 LaNi$_{4.6}$Cu$_{0.4}$; 3 MnNi$_{4.6}$Fe$_{0.4}$; 120 g each	20	2	80	56	0.3	2	No data	Nat. Inst. for R&D of Isotopic and Molecular Technologies (RO); water heating/cooling	[183]
2010	Periodic	2	1	1 La$_{0.5}$Ce$_{0.5}$Ni$_{4.5}$Mn$_{0.5}$Al$_{0.05}$; 2 Ti$_{0.8}$Zr$_{0.2}$Co$_{0.5}$Fe$_{0.5}$Ni$_{1.2}$; 7 kg in total	25	50	150	700	2	60	No data	Zhejiang University (CN); oil heating/cooling	[184]
2011	Periodic	1	3	LaNi$_5$, Ca$_{0.6}$Mm$_{0.4}$Ni$_5$, Ca$_{0.7}$Mm$_{0.3}$Ni$_5$	10	13–40	90	100–150	No data			Joint US–KR team; water heating/cooling MH compacts of Cu-encapsulated IMC particles (5/3 g) with Sn binder (0.5/0.3 g), 170–200 bar compacting pressure	[158]
		2	3	1 LaNi$_5$; 2 LaNi$_5$, Ca$_{0.6}$Mm$_{0.4}$Ni$_5$, Ca$_{0.7}$Mm$_{0.3}$Ni$_5$	10	7	125	100–160					
2012	Continuous	2	6	1 LaNi$_5$ (6 × 14 kg); 2 La$_{0.5}$Ce$_{0.5}$Ni$_5$ (6 × 10 kg)	10–15	2–5	150	150–160	15	10	No data	Russian Acad. Sci.: Spec. Design & Engineering Bureau in Electrochemistry (RU); water cooling, steam heating;	[58]
2012	Continuous	2	2	1 (La,Ce)Ni$_5$ (2 × 15 kg); 2 (Ti,Zr)(Fe,Mn,Cr,Ni)$_2$ (2 × 12 kg)	20	10	120	200	1	30	1.65	Univ. Western Cape (ZA); water cooling, steam or overheated water heating;	[94]
2012	Periodic	1	1	1 La$_{0.9}$Y$_{0.1}$Ni$_5$Al$_{0.3}$ (594 g); 2 V/LaNi$_5$ (594 g)	20	20	175	350	0.19	6	No data	Inst. of Refrigeration and Cryogenics Eng., Shanghai Jiaotong Univ. (CN); water cooling, water/oil? heating	[75]
		2	1		2080?	20	95/75?	380	0.28				
2012	Continuous	2	1	1 AB$_5$; 2 AB$_5$	50	160	190	600	No data	40	No data	Univ. of Birmingham (UK); oil heating, water cooling;	[185] and [186]
2013	Continuous	2	3	1 AB$_2$; 2 AB$_5$	30	10–30	120	200	5–10	No data		HYSTORSYS AS (NO); oil heating/cooling;	[187] and [188]

[a] Adopted for pumping hydrogen and tritium mixture by pressure transmission via mercury U-tube.

[b]86 °C at the output, the value of ΔTH was used for the estimation of the consumed heat for efficiency calculations.

[c]The efficiency has been calculated by the authors of this review starting from the performance data.

[d]Up to 4000 bar using the second, cryogenic stage.

[e]Planned.

[f]As presented in the original works, most probably, this is % of Carnot efficiency (19.3% for the specified temperature range).

[g]Miniature hydride heat exchangers retain hydride alloy within 1/16″ OD Tubes.

[h]Stage 1 – capacity 2000 L, Stage 2 – capacity 1000 L. 300–400 bar at 1st stage (separate collection to the receiver).

[i]For stages 1/2.

Main design features of the MH compressors include type(s) of the used hydride-forming material(s), mode of operation (periodic or continuous), as well as number of compression stages.

Integration of MH containers for H_2 compression (Section 3.3.1) into compressor assemblies realises one of the general layouts schematically shown in Fig. 7. The corresponding engineering solutions are mainly related to (i) selection of number of stages and proper MH material(s) allowing to achieve the required H_2 compression in the available heating/cooling temperature range; (ii) design of the gas-distributing system; (iii) management of periodic heating and cooling of the MH containers; (iv) control and automation of operation; and (v) solutions aimed at the increase of compressor's efficiency.

As a first step of the selection of MH material, the thermodynamic approach described in Section 2.1 can be used. It should be noted that the selection has to foresee a certain margin in the material's performances, i.e. for the compression stage operating between TL and TH, the value of PL should be lower and the value of PH higher than H_2 suction and discharge pressures, respectively. This will provide a driving force necessary to achieve acceptable rates of the H absorption and desorption processes. Further evaluation of charge/discharge dynamic performances of the selected design of MH container (see previous section) will allow one to determine required number of the containers in an assembly, and to estimate important performance characteristics, like cycle productivity, cycle time, consumption of power or heating/cooling fluids, etc.

One-stage periodically operated MH compressors usually comprise one (Fig. 7(a)) or several connected in parallel MH containers/compression elements. They normally have a simple gas-distributing system on the basis of shut-off valves which provides connection of a gas manifold of a cooled compression element to low-pressure H_2 input and, when the compression element is heated up, to high-pressure H_2 output port of the compressor. Due to its simplicity and flexibility, this solution found numerous applications (see Table 2), mainly in laboratory practice. In combination with electric heating to moderately high temperatures (~500 °C), it can provide generation of very high hydrogen pressures, up to 5 kbar [179]; another option to generate kilobar-range H_2 pressures is in a combination of MH compression (first stage, up to 400 bar) with the cryogenic cooling–heating cycle (second stage) [176].

In a multistage periodically-operated MH compressors gas manifolds of the stages are connected sequentially in the ascending order; opening and closing of the valves is synchronised with the alternate periodic heating and cooling of the compression elements, e.g. stages 1, 3   stage 2 for the three-stage compressor (see example in Ref. [138]).

The gas-distributing systems of continuously-operating MH compressors provide switching of H_2 flow passing from a low-pressure input port to gas manifold of the cooled compression element(s) of the first stage, further from gas manifolds of the heated compression elements to the cooled compression elements of the next stage, and, finally, from gas manifold of the heated compression elements of the last stage to the high-pressure H_2 output port (see Fig. 7(b, c)). The switching can be provided by manual or remotely actuated shut-off valves whose operation must be synchronised with periodic heating and cooling of the compression elements. A commonly-used solution which allows to simplify the compressor assembly and to reduce its cost is in the usage of one-way (check) valves (Fig. 8) which automatically provide flowing of H_2 from higher pressure manifolds to lower pressure ones, so that the operation of the compressor can be achieved only by thermal management (see, e.g. Refs. [100], [101] and [102]). The check valve solution is similar to the one conventionally applied in mechanical compressors. However, the heat-driven MH compressors significantly differ from their mechanical analogs by much slower rate of pressure increase/decrease when passing from charge to discharge mode and vice versa. Typical duration of the charge/discharge cycle in MH hydrogen compressors varies from ~1 to ≥30 min that is much longer than the duration of the conventional mechanical compression cycle (≪1 s). This fact increases the probability of malfunction of the check valves resulting in an H_2 backflow. Thus, when introducing check valves in gas-distributing system of MH compressor assembly, special attention has to be paid to the measures decreasing a probability of the backflow. According to the authors' experience, the problem can be addressed by a proper selection of the check valves (non-rotating stem and high enough cracking pressure), as well as by elimination of possibility of gas contamination by the MH powder. Bowman et al.[193] have conducted extensive cycling tests of Nupro "CW-type" check valves showing no degradation after over 43,000 cycles as well as evaluated porous stainless steel filters to retain hydride powder while allowing sufficient H_2 flow rates.

Suction/H absorption mode of the MH containers/compression elements is provided by the cooling using natural [66], [139] and [179] or forced [133] and [141] air convection, or flow of cooling fluid (water [62],[148] and [172] or oil [181], [182], [184] and [187]). Some solutions envisage the cooling using a chiller plate[142], or thermoelectric/Peltier modules [96]. To provide high-pressure hydrogen discharge, the MH containers are heated up using electric heaters [66], [71], [128], [133], [139], [141], [142], [173], [174],[175], [176] and [179], thermoelectric modules [96], or flow of a heating fluid (hot water at $TH < 100$ °C [62],[135], [136], [138], [148], [151], [170], [171], [177], [180], [181] and [183], oil [181], [182], [184], [185], [186],[187] and [188], overheated water [94] and [158] or steam [58], [94] and [153] at the higher temperatures). One solution by Ergenics uses heating of the MH containers by the burning of natural gas [134].

The operation of permanently operating MH compressors is usually controlled by time relays which provide a periodic switching of the MH containers between suction/absorption and discharge/desorption modes. As a rule, both absorption and desorption time setpoints are the same that allows to simplify system layout. At the same time, it was noted that in most cases the desorption time is shorter than the absorption one[108], [118] and [126].

It has to be noted that the operating parameters of the MH compressors mostly influence their productivity while the working pressure–temperature ranges are mainly determined by the thermodynamic properties of the selected MH material(s) (see section 2.1). First of all, the productivity depends on the variations of the H_2 suction pressure, cooling temperature, cycle time, and, in a lesser extent, heating temperature and H_2 discharge pressure [94] and [108]. This influence is especially pronounced in a multistage layout when the combination of the factors specified above mainly affects on hydrogen flow rate between the stages which often becomes a step limiting the total productivity of the compressor [94].

The importance of heat recovery for the increase of MH compressors efficiency was underlined in a number of studies, see, e.g. Refs. [104] and [106]; the corresponding engineering solutions can be found in patents [189], [190], [191] and [192]. However, only few system developments known to the authors realise this approach. An attempt to apply the heat recovery by the circulation of water between hot and cold MH containers/compression elements after completion of H_2 absorption–desorption cycle was undertaken by South African co-authors of the present review [94] and [97]. The solution was shown to be feasible; moreover, its application provided more stable operation of the compressor using water for the cooling and steam for the heating. At the same time, the introduction of the additional circulation loop results in the complication of the system layout and in the increase of its cost. It also results in the decrease of the system productivity.

Examples of medium-to-large scale permanently operated MH compressors are presented in Fig. 17.

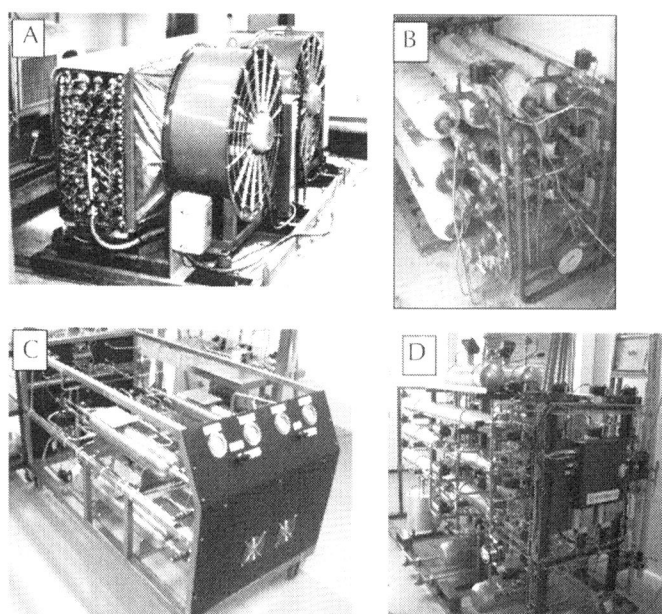

Figure 17: Medium-to-large scale MH compressors: A – Institute for Mechanical Engineering Problems of the National Academy of Sciences of Ukraine (1998, 3–150 bar/10 m³/h) [141]; B – Institute of Problems of Chemical Physics/Russian Academy of Science, Special Design Engineering Bureau in Electrochemistry, Russia (2012, 2–160 bar/15 m³/h) [58]; C – South African Institute for Advanced Materials Chemistry/University of the Western Cape (2012, 10–200 bar/1 m³/h) [94]; D – HYSTORSYS AS, Norway (2013, 10–200 bar/10 m³/h) [188].

Applications of MH H$_2$ Compressors

Since Reilly et al. [62] described a hydride compressor that used VHx in 1971, a number of possible applications has proposed. Some examples are cited by Sandrock [6] and [68], Dantzer [7], Bowman and Fultz [11]. However, the most diverse collection of hydride compressors and potential applications can be seen in the literature by Ergenics [81], [98], [99], [100], [101], [102], [133], [134] and [172].

This section briefly presents the most important applications of MH compressors including historical summary, and an overview of the recent developments.

Isotope Handling

Metal hydrides have been used internationally in the research laboratories, nuclear energy and defence industries for decades to store and process hydrogen isotopes, protium, deuterium, and tritium[194] and [195]. Prior to 1970 the binary hydrides of titanium, zirconium, palladium, and uranium were only utilised [196]. Often these metal hydrides served concurrent roles of collecting, storing, purifying, transporting, and isotope separation rather than to serve as explicit compression applications. However, several organisations in the U.S. Nuclear Defence industry that included Los Alamos Scientific Laboratory (Alamos NM) [197], the Mound Laboratory (Miamisburg OH) [196] and the Savannah River Site (Aiken SC)[198] generated and supplied highly purified tritium gas at pressures of several bar (typically <5 bar) by heating storage beds of uranium powder to circa 650 K and higher temperatures that can be regarded as a single step MH compressor. There were issues of tritium inventory control and managements due to enhanced permeation of this radioactive gas through the stainless steel bed walls at these elevated temperatures [198] and [199]. In the mid-1980s, the Savannah River Site (SRS) started the development[200], [201] and [202] of an enhanced tritium processing facility where several metal hydrides based upon the AB_5 and AB_2 alloys were employed in various roles including as compressors to replace conventional mechanical compressor technology. Using two or three stage compression and different alloys, compression of hydrogen isotopes up to the pressures of 620 bar were achieved for maximum bed temperatures of around 460 K [203]. A photograph of an example 3-stage compressor built at SRS is shown in Fig. 18 where the alloys were $LaNi_{4.5}Al_{0.5}$ (Stage-1 to 14 bar), $LaNi_{4.9}Al_{0.1}$ (Stage-2 to 200 bar), and $TiCr_{1.8}$ (Stage-3 to 620 bar). These compressors provided safe and reliable operation during more than 20 years of their use [203].

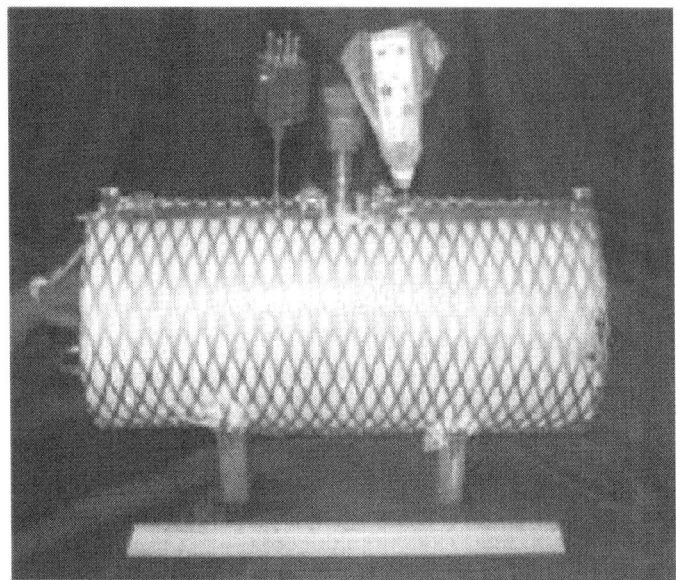

Figure 18: A 3-stage metal hydride (LaNi$_{4.5}$Al$_{0.5}$, LaNi$_{4.9}$Al$_{0.1}$, and TiCr$_{1.8}$) compressor fabricated at the tritium facility of the Savannah River Site (Aiken SC USA) for compression of hydrogen isotopes to pressures of ~620 bar.

Similar development which uses one-stage MH compression for the supply of hydrogen isotopes (hydrogen, deuterium, tritium, or their mixture) at very high, up to 5 kbar, pressures was reported by Russian Research Institute of Experimental Physics. The periodically operated compression system uses decomposition of vanadium dihydride at $T \leq 360$ °C. It is intended for the use in various experimental studies, including muon-catalysed fusion [179].

Cryogenics/Space

In 1972, van Mal was the first to report that metal hydride compressors could be used to form liquid hydrogen via Joule–Thomson (J–T) expansion [204] and [205]. Subsequently, a number of laboratory demonstrations of hydrogen liquefaction using metal hydride compression were done as previously reviewed by Bowman [8] and [9] and presented in a number of original developments, see, e.g. Refs. [206],[207] and [208].

A 3-stage hydride compressor was developed and built by a team at NASA Jet Propulsion Laboratory (JPL) and Aerojet Electronic Systems Division (Azusa CA USA) that periodically formed solid molecular hydrogen (s-H_2) at temperatures below 10 K from 100 bar H_2 gas [173] when integrated with a J–T capillary tube and three Stirling cryocoolers operating at $T > 60$ K [174]. The overall H_2 compression ratio from these hydride beds was 8.3×10^5. The hydride compressor assembly is shown in Fig. 19. This cryocooler was operated in earth orbit on-board the Space Shuttle Orbiter Endeavour during May 1996 and successfully generated s-H_2 at 10.4 K on its first cool down cycle [175]. However, small metallic particles that were free floating in the storage volume at zero gravity were swept into control valve preventing J–T valve from fully sealing during subsequent space flight tests [175]. Consequently, only liquid hydrogen at the temperature 18–21 K could be obtained. When these particles were removed from the damaged valve seat at JPL following the space flight, the repaired cryocooler operated normally and was again able to generate s-H_2 at temperatures between 9.4 K and 10.0 K during post-flight tests in the laboratory [175].

Figure 19: JPL 10 K sorption cryocooler hydride compressor bed assembly. (1) Fast absorption hydride bed (LaNi$_{4.8}$Sn$_{0.2}$); (2) low pressure hydride bed (ZrNi); and (3) high pressure hydride bed (LaNi$_{4.8}$Sn$_{0.2}$).

Starting from 1997, JPL developed and fabricated two completely redundant 20 K sorption cryocoolers for the European Space Agency (ESA) mission Planck [209] aiming at mapping and measuring the Cosmic Microwave Background (CMB) radiation with higher resolution and sensitivity than any prior study. Comprehensive descriptions of the Planck Mission and the thermal control systems for its satellite are given in Refs. [209] and [210], respectively. Details on the development and prior-to-launch testing of the sorption cryocoolers and metal hydride (i.e., $LaNi_{4.78}Sn_{0.22}Hx$) compressors are available in Refs.[142] and [143] and the various papers cited therein. The Planck satellite was launched in May 2009 with the two sorption cryocooler operated in a serial fashion. Excessive degradation in performance of the first unit was noted after less than one year of flight operation and it was switched off. The second flight unit operated within required performance levels for over three years until completion of the Planck flight mission in October 2013. Reasons for this difference are not yet identified, but the extensive flight data files are being reviewed to detect potential causes. Fig. 20 displays the two compressor assemblies being integrated on the support mounting of the Planck satellite. The individual metal hydride beds are mounted onto radiator plates on the outer panels of the satellite. An overview of the performance of the sorption cryocoolers during the first year of space flight operation is given by Ade et al. [211].

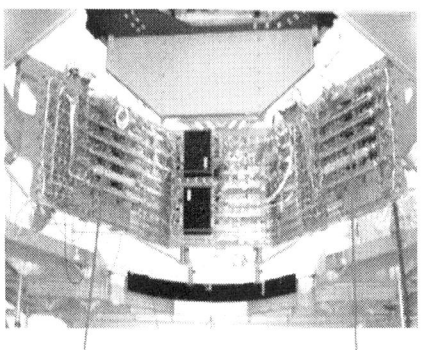

Hydride Compressor Assemblies

Figure 20: Photograph taken during the integration of the hydrogen sorption cryocoolers onto the support structure of the Planck satellite.

Utilisation of Low-Grade Heat

As it was already mentioned, the main advantage of MH hydrogen compressors is in the conversion of waste heat ($T < 200$ °C) into the energy of compressed hydrogen which can be further utilised.

In 1979 Nomura et al. [170] developed and successfully tested a piston engine which used one stage MH compressor on the basis of $LaNi_5$ and operated at 20–80 °C providing efficiency of energy conversion of 7.7%, or about 50% of the Carnot efficiency. A year later the first prototype of MH-based ($LaNi_{4.63}Al_{0.37}$) water pump operating in the same temperature range was developed at Sandia National Laboratories[171]. Use of MH compressors for water pumping driven by solar heat was intensively studied in the early 2000s [212], [213] and [214]. The systems were shown to be promising in distributed stand-alone applications capable of daily pumping up to 3000 L of water over a height of 15 m using 1 m^2 solar collector area [214].

Demonstration of a metal hydride heat engine developed by Ergenics was able to convert the heat from e.g. solar hot water into electricity and can be found in Ref. [215]. The issues of upscaling similar solutions, by integration of MH compressor with radial-axial turbine expansion engine were recently considered by Rusanov et al. [216]. Analysis of performances of energy conversion using MH hydrogen compression in comparison with conventional organic Rankine cycle [217] showed that the MH compression is more promising, due to lower operation cost, higher exergy efficiency and thermal COP, higher output power, and acceptable capital costs. The advantages are especially pronounced for the utilisation of waste industrial heat.

Thermally Driven Actuators

Developments of pneumatic actuators on the basis of MH hydrogen compression were considered in earlier reviews of MH applications [6] and [68]. Their advantages include compactness, ability to develop high forces, smooth actuation, silent and vibration-free action, simplicity in design and operation. Since the late 1980s Japan Steel Works has carried out intensive R&D of the MH actuators, which were used in various types of rehabilitation equipment [218], [219] and [220]. As a rule, the actuators are driven by thermoelectric/Peltier elements used for heating and cooling of the MH; pressure transmission from compressed H_2 to mechanical or hydraulic actuating mechanism is provided by bellows. The developed devices have an impressive performance: for example, use of 12 g of $(Ca,Mm)(Ni,Al)_5$ alloy allows to lift 50 kg load by 5 cm [220].

The data about development of MH actuators over last decade can be found in Refs. [220], [221], [222],[223] and [224] and references therein. The developments are mainly focused on the integration of MH with pneumatic McKibben actuators/"artificial muscles" [222] and [223], as well as study of feasibility of usage of low-grade (e.g. solar) heat to drive the actuators [224].

Hydrogen Refuelling Stations

Hydrogen refuelling infrastructure takes a significant part of the capital investments for the introducing fuel cell powered vehicles and must be taken into account in the assessment of their economic feasibilities. Despite a certain number of hydrogen refuelling stations operating worldwide, they are not introduced broadly enough, mainly because of their high costs ranging between $500,000 and $5,000,000 per installation [225]. The most expensive H_2 refuelling components originate from: (i) on-site hydrogen production and (ii) hydrogen compression. According to techno-economic analysis presented in Ref. [226], the contribution of hydrogen compression to the total station cost is about 20%.

Cost–performance optimisation of the H_2 refuelling infrastructure can be achieved by the improvement of hydrogen compression technology. A promising way for that is the application of thermally-driven metal hydride hydrogen compressors characterised by simplicity in design and operation, reliability and minimum maintenance, with potentially low price and ability to utilise waste heat, instead of electricity, for the H_2 compression. An example of the integration of MH compressor in hydrogen refuelling infrastructure is presented in patent [227].

Fig. 21 presents features of the HyNor Lillestrøm hydrogen refuelling station in Norway [228] and [229]. The station uses on-site produced hydrogen and has three H_2 storage tanks (200, 450 and 1000 bar) on the basis of high-pressure composite cylinders. H_2 compression is carried out using metal hydride (seeFig. 17(D)) and mechanical (membrane) hydrogen compressors. Hydrogen is dispensed to the FC vehicles at the pressure of 700 bar.

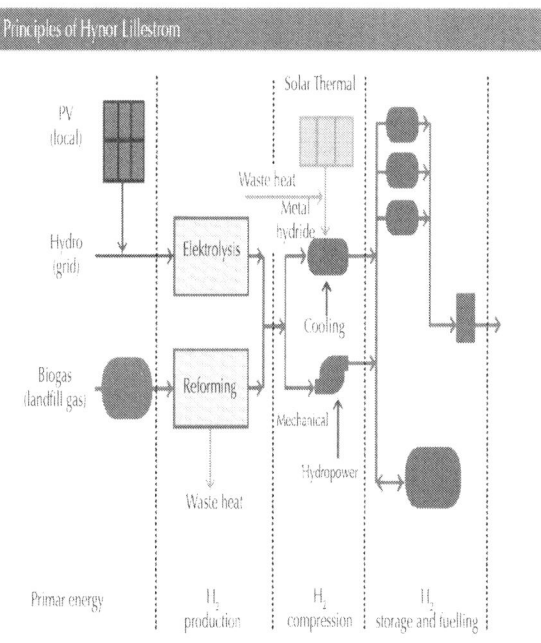

Figure 21: Schematics of HyNor Lillestrøm hydrogen refuelling station for FC vehicles.

Economic Estimations

For understandable reasons, the available information about prices for MH hydrogen compressors is scarce. However, there are indications that even the prototype costs for medium-to-large scale MH compressors can be comparable with the prices for commercially available mechanical compressors. So, the estimated cost of the 10–100 bar/0.42 m³/h prototype built in Brazil [71] was reported as $23,000 versus $27,000 for PPI mechanical compressor (1989). The cost of 3–150 bar/10 m³/h MH compressor built in Ukraine in 1997 (Fig. 17(A); [141]) was estimated as $32,500–$39,000; the investments could be returned in 5–6 months, due to high price difference between low- and high-pressure hydrogen. The capital costs for 0.5–430 bar

MH compressor estimated by Linde North America, MRT and Ergenics in 2009 [136]were about $66,000.

Fig. 22 presents costs breakdown (in %) for the prototype MH compressors recently developed with participation of South African and Norwegian co-authors of this review. It can be seen that the most expensive part of the compressor is metal hydride containers/ compression elements, and optimisation of their design and manufacturing technology could result in the significant price decrease.

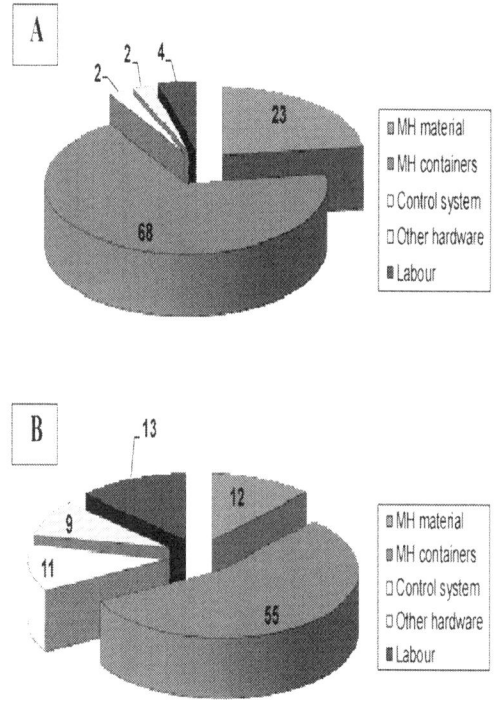

Figure 22: Cost breakdown (in %) for prototype MH compressors: A: 10–200 bar/1 m³/h, South Africa (Fig. 17C); B – 10–200 bar/10 m³/h, Norway (Fig. 17D; the data were provided by Dr. Jon Eriksen, HYSTORSYS AS). The labour costs relate to system integration only.

It is expected that maintenance cost for the MH compressors will be significantly lower than for their mechanical analogues. Thus, the implementation of MH compressors has clear economic advantages.

CONCLUDING REMARKS

Analysis of the reference data summarised in the present review shows a stable growth of the R&D activities in the development of the metal hydride hydrogen compression technology as illustrated by Fig. 23(A) where the total numbers of bibliographic entries explicitly focused on the MH H$_2$ compression technology are shown. Significant intensification of the R&D on MH hydrogen compressors was observed recently (after 2008) making the present review very timely.

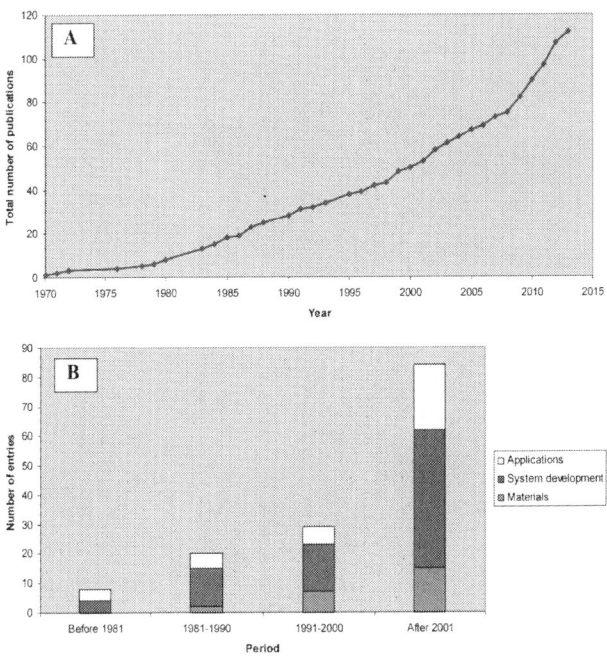

Figure 23: Analysis of the reference data on the MH compressors: A – number of publications and patents; B – scope of the work.

While early works (i.e., during the 1970s–1980s) mainly dealt with the proof of concept and outlining possible applications of the MH compressors on the basis of general features of hydrogen–metal systems, further R&D in this field became more focused, including special research of the MH compression alloys, as well as optimisation of system layout towards improvement of its performances (extension of the operating pressure range, increase of productivity and efficiency, etc.). Many works, especially those published in the last decade, are aimed at the alignment of the material and system features with the requirements of specific applications (Fig. 23(B)[4]).

Commercial competitiveness of the metal hydride compression in comparison with alternative hydrogen compression technologies is justified by both technical and economic considerations.

Use of the waste industrial heat is a major winning argument for use of the MH compression to dramatically decrease its operational costs. Furthermore, costs of the selected MH materials and their availability have a primary importance for a broad application of the technology.

Efficient metal hydride compression process requires

-a high compression ratio (small slope of the isotherms, low hysteresis and appropriate thermodynamics of the metal–hydrogen system);

-high productivity and efficiency (low number of compressions steps, fast kinetics of hydrogen exchange, efficient heat transfer, low transient heat losses);

-long and reliable operation (high cycle stability of metal hydride materials at the operating conditions, efficient system design).

Optimisation of the performances of the MH compressors strongly depends on finding a compromise between a number of contradicting factors, which could be divided into three groups (Table 3).

Table 3: Factors influencing the performances of metal hydride hydrogen compressors

Group	Factor	Performances				
		Suction pressure	Discharge pressure	Efficiency	Productivity	Safety and reliability
A Properties orklll	Entropy of hydride formation. !AS˚1	Strong increase	Strong increase			
	Enthalpy of hydride formation, lAll˚1			Increase Increase	Strong In-crease	
	Reversible H sorption capacity, AC					
	Hysteresis, er	Increase	Decrease	Decrease		
	altos > Heat capac-ity, c		Decrease (for high pressures)	Decrease	Decrease	
	Dilatation, AV/Vo			Decrease	Decrease	Increase
	Sorption / desorption kinetics, K,				Increase	

B Design / technology	Overall heat transfer coefficient, K_m			Increase	Strong increase	
	Overall mass transfer coefficient, K_a				Increase	
	Dead space, G			Decrease	Decrease	Strong increase
	Material consumption, G			Decrease	Increase	Strong increase
	Heat recovery efficiency, a			Increase	Decrease	
	Number of stages, n	Decrease	Increase	Decrease	Decrease	Decrease
C' Operating condition,	Lower temperature level, T_t			Decrease	Decrease	
	Upper temperature level, $1''_{,,}$		Strong increase	Increase	Increase	
	Cycle duration, tc	Decrease	Increase	Increase		Increase

The first group (A) concerns the development of MH materials for the H_2 compression. These should assure matching specified pressure and available temperature ranges, having maximal reversible sorption capacity, minimal dilatation and heat capacity, in addition to the factors mentioned above in this section. Furthermore, cyclic stability and poisoning resistance of an MH alloy should be taken into account.

The second (B) and third (C) groups of factors are related to the optimisation of the design of the MH compressor, technology of its manufacturing, and operation conditions (i.e. addressing and resolving the engineering problems without adversely impacting system mass/volume values, component manufacturability or assembly, or reliability during long term temperature/pressure cycling).

Finally, successful implementation of metal hydride hydrogen compressors is strongly dependent on their manufacturing costs of which the most expensive items are metal hydride containers/compression elements. Optimisation of the manufacturing will result in the cost reduction and, accordingly, will secure the commercial success of the metal hydride hydrogen compression technology.

ACKNOWLEDGEMENTS

This invited review written on a request of the Editor-in-Chief of International Journal of Hydrogen Energy, Emre A. Veziro lu, summarises both individual and joint efforts of the co-authors for more than three decades.

Development and characterisation of MH materials for hydrogen compression was done via Program of Research Co-operation between Norway and South Africa funded by Research Council of Norway (RCN) and NRF in South Africa (2007–2010, Project #180344).

Volodymyr A. Yartys greatly acknowledges financial support received from the Research Council of Norway (projects 191106 "Thermally Managed Systems for Storage, Compression and Supply of Hydrogen Gas" and 180200 "Hybrid Hydrogen Storage Solutions"), Hystorsys AS and CMR Prototech. He would like to sincerely thank colleagues from IFE (Dr. Jan Petter Mæhlen), Hystorsys AS (Dr. Jon Eriksen, Dr. Roman V. Denys and Mr. Christoph Cloed) and CMR Prototech (Mr. Arild Vik) for the long standing fruitful collaboration on materials science studies of metal hydrides and on the technologies of hydrogen storage and compression.

Robert C. Bowman, Jr. thanks the Fuel Cell Technology Office of the U.S. Department of Energy, Office of Energy Efficiency and Renewable Energy for their support of his work at the Oak Ridge National Laboratory. He also thanks Dr. Ted Motyka for providing information on the hydride compressor technology at the Savannah River Site and Prof. Ted B. Flanagan for more than two decades of fruitful collaborations on properties of metal hydrides. He also greatly appreciates the contributions made by his numerous colleagues at the Jet Propulsion Laboratory during the development of the hydrogen sorption cryocoolers. This manuscript has been authored by UT-Battelle, LLC, under Contract No. DE-AC05-00OR22725 with the U.S. Department of Energy. The United States Government retains and the publisher, by accepting the article for publication, acknowledges that the United States Government retains a non-exclusive, paid-up, irrevocable, world-wide license to publish or reproduce the published form of this manuscript, or allow others to do so, for United States Government purposes.

Mykhaylo Lototskyy and Bruno G. Pollet, would like to acknowledge support of Eskom Holdings Ltd (South Africa) who mainly funded the developments of MH hydrogen compressors at South African Institute for Advanced Materials Chemistry (SAIAMC). The invaluable contribution of the Department of Science and Technology (DST) in South Africa within a number of Research, Development and Innovation programmes and projects, including Hydrogen South Africa National Flagship Programme (HySA), is greatly acknowledged. The R&D activities in the development and optimisation of MH hydrogen compressors were also supported by Impala Platinum Ltd (South Africa). Investments from the industrial funders have been leveraged through the Technology and Human Resources for Industry Programme, jointly managed by the South African National Research Foundation and the Department of Trade and Industry (NRF/DTI; THRIP projects TP2010071200039, TP2011070800020, and TP1207254249). Mykhaylo Lototskyy also acknowledges NRF support via incentive funding grant 76735.

Finally, Mykhaylo Lototskyy would like to thank a metal hydride team from the Institute for Mechanical Engineering Problems of the National Academy of Sciences of Ukraine (now Hydrogen Power Engineering Department), where he was working for 20 years and contributed to the generation of numerous R&D outputs part of which was analysed in this review. Sincere personal gratitude is addressed to the team leader, Professor V.V.Solovey, as well as to the memory of late team members, Yuri Shmal›ko (1948–2008) and Alexander Ivanovsky (1954–2013).

REFERENCES

1. Lynch JF, Maeland AJ, Libowitz GG. Hydrogen compression by metal hydrides. In: Veziroglu TN, Taylor IB, editors. Hydrogen energy progress V. Proc. 5th world hydrogen energy conference, vol. 1. Oxford: Pergamon Press; 1984. pp. 1327e37.

2. Dantzer P, Orgaz E. Thermodynamics of the hydride chemical heat pump: model (I). J Chem Phys 1986;85:2961e73.

3.   Dantzer P, Orgaz E. Thermodynamics of hydride chemical heat pump: II. How to select a pair of alloys. Int J Hydrogen Energy 1986;11:797e806.

4.   Orgaz E, Dantzer P. Thermodynamics of the hydride chemical heat pump: III. Considerations for multistage operation. J Less-Common Met 1987;131:385e98.

5.   Solovey VV. Metal hydride thermal power installations. In: Veziroglu TN, Protsenko AN, editors. Hydrogen energy progress VII. Proc. 7th world hydrogen energy conference, vol. 2. Oxford: Pergamon Press; 1988. pp. 1391e9.

6.   Sandrock G. Applications of hydrides. In: Yurum Y, editor. Hydrogen energy system. Production and utilization of hydrogen and future aspects. Kluwer Acad Publ. NATO ASI Ser. Ser E, 295; 1995. pp. 253e80.

7.   Dantzer P. Metal-hydride technology: a critical review. In: Wipf H, editor. Hydrogen in metals III. Properties and applications. Berlin-Heidelberg: Springer-Verlag; 1997. pp. 279e340.

8.   Bowman Jr RC, Kiehl B, Marquardt E. Closed-cycle JouleThomson cryocoolers. In: Donabedian M, editor. Spacecraft thermal control handbook. Cryogenics, vol. 2. El Segundo, CA, USA: The Aerospace Press; 2003. pp. 187e216.

9.   Bowman Jr RC. Development of metal hydride beds for sorption cryocoolers in space applications. J Alloys Compds 2003;356e357:789e93.

10. Muthukumar P, Groll M. Erratum to "metal hydride based heating and cooling systems: a review" [International Journal of Hydrogen Energy (2010) 35: 3817e3831]. Int J Hydrogen Energy 2010;35. 8816e8829*. * The authors [10] recommend to refer to the Erratum since the original article contains several publisher's errors corrected in the second version.

11. Bowman RC, Fultz B. Metallic hydrides I: hydrogen storage and other gas-phase applications. MRS Bull 2002;27(9):688e93.

12. Tarasov BP, Lototskii MV, Yartys' VA. Problem of hydrogen storage and prospective uses of hydrides for hydrogen accumulation. Russ J Gen Chem 2007;77(4):694e711.

13. Flanagan TB, Oates WA. Thermodynamics of intermetallic compoundehydrogen systems. In: Schlapbach L, editor. Hydrogen

in intermetallic compounds. I. Electronic, thermodynamic and crystallographic properties, preparation. BerlineHeidelberg; 1988. pp. 49e85.

14. Sandrock G. Hydrogenemetal systems. In: Yurum Y, editor. Hydrogen energy system. Production and utilization of hydrogen and future aspects. Kluwer Acad Publ. NATO ASI Ser. Ser E, 295; 1995. pp. 135e66.

15. Lototsky MV, Yartys VA, Zavaliy IYu. Vanadium-based BCC alloys: phase-structural characteristics and hydrogen sorption properties. J Alloys Compds 2005;404e406:421e6.

16. Kodama T. The thermodynamic parameters for the LaNi5xAlxeH and MmNi5xAlxeH systems. J Alloys Compds 1999;289:207e12.

17. Khyzhun OYu, Lototsky MV, Riabov AB, Rosenkilde C, Yartys VA, Jørgensen S, et al. Sn-containing (La,Mm) Ni5xSnxH5e6 intermetallic hydrides: thermodynamic, structural and kinetic properties. J Alloys Compds 2003;356e357:773e8.

18. Dehouche Z, Grimard N, Laurencelle F, Goyette J, Bose TK. Hydride alloys properties investigations for hydrogen sorption compressor. J Alloys Compds 2005;399:224e36.

19. Luo S, Luo W, Clewley JD, Flanagan TB, Bowman RC. Thermodynamic and degradation studies of LaNi4.8Sn0.2eH using isotherms and calorimetry. J Alloys Compds 1995;231:473e8.

20. Dantzer P, Meuner F. What materials to use in hydride chemical heat pumps. Mat Sci Forum 1988;31:1e18.

21. Diaz H, Percheron-Guegan A, Achard JC. Thermodynamic and structural properties of LaNi5yAly compounds and their related hydrides. Int J Hydrogen Energy 1979;4:445e54.

22. Singh RK, Gupta BK, Lototsky MV, Srivastava ON. On the synthesis and hydrogenation behaviour of MmNi5xFex alloys and computer simulation of their PeCeT curves. J Alloys Compds 2004;373:208e13.

23. Libowitz GG, Maeland AJ. Use of vanadium-based solid solution alloys in metal hydride heat pumps. J LessCommon Met 1987;131:275e82.

24. Huston EL, Sandrock GD. Engineering properties of metal hydrides. J Less-Common Met 1985;74:435e43.

25. Luo G, Chen JP, Li SL, Chen W, Han XB, Chen DM, et al. Properties of La0.2Y0.8Ni5xMnx alloys for high-pressure hydrogen compressor. Int J Hydrogen Energy 2010;35:8262e7.

26. Kapischke J, Hapke J. Measurement of the pressurecomposition isotherms of high-temperature and lowtemperature metal hydrides. Exper Therm Fluid Sci 1998;18:70e81.

27. Guo X, Wang S, Liu X, Li Z, Lu¨ F, Mi J, et al. Laves phase hydrogen storage alloys for super-high-pressure metal hydride hydrogen compressors. Rare Met 2011;30(3):227e31.

28. Wang X, Chen R, Zhang Y, Chen C, Wang Q. Hydrogen storage properties of (LaeCeeCa)Ni5 alloys and application for hydrogen compression. Mater Lett 2007;61:1101e4.

29. Au M, Wang Q. Rare earth-nickel alloy for hydrogen compression. J Alloys Compds 1993;201:115e9.

30. Zotov TA, Sivov RB, Movlaev EA, Mitrokhin SV, Verbetsky VN. IMC hydrides with high hydrogen dissociation pressure. J Alloys Compds 2011;509S:S839e43.

31. Johnson JR. Reaction of hydrogen with the high temperature (C14) form of TiCr2. J Less-Common Met 1980;73:345e54.

32. Wang X, Chen R, Zhang Y, Chen C, Wang Q. Hydrogen storage alloys for high-pressure suprapure hydrogen compressor. J Alloys Compds 2006;420:322e5.

33. Brodowsky H, Yasuda K, Itagaki K. From partition function to phase diagramstatistical thermodynamics of the LaNi5eH system. Z Phys Chem 1993;179:45e55.

34. Beeri O, Cohen D, Gavra Z, Johnson JR, Mintz MH. Thermodynamic characterization and statistical thermodynamics of the TiCrMneH (D) system. J Alloys Compds 2000;299:217e26.

35. Shilov AL, Efremenko NE. Effect of sloping pressure "plateau" in two-phase regions of hydride systems. Russ J Phys Chem 1986;60:3024e8.

36. Larsen JW, Livesay BR. Hydriding kinetics of SmCo5. J LessCommon Met 1980;73:79e88.

37. Fujitani S, Nakamura H, Furukawa A, Nasako K, Satoh K, Imoto T, et al. A method for numerical expressions of P-C isotherms of hydrogen-absorbing alloys. Z Phys Chem 1993;179:27e33.

38. Lototsky MV. A modification of the LachereKierstead theory for simulation of PCT diagrams of real "hydrogenehydrideforming material" systems. Kharkov Univ Bull/No. 477: Chem Ser 2000;5(28):45e53.

39. Lototsky MV, Yartys VA, Marinin VS, Lototsky NM. Modelling of phase equilibria in metalehydrogen systems. J Alloys Compds 2003;356e357:27e31.

40. Park CN, Luo S, Flanagan TB. Analysis of sloping plateaux in alloys and intermetallic hydrides. I. Diagnostic features. J Alloys Compds 2004;384:203e7.

41. Lacher JR. A theoretical formula for the solubility of hydrogen in palladium. Proc Roy Soc (Lond) 1937;A161:525e45.

42. Kierstead HA. A theory of multiplateau hydrogen absorption isotherms. J Less-Common Met 1980;71:303e9.

43. Flanagan TB, Clewley JD. Hysteresis in metal hydrides. J Less-Common Met 1982;83:127e41.

44. Balasubramaniam R. Hysteresis in metalehydrogen systems. J Alloys Compds 1997;253e254:203e6.

45. Murthy SS. Heat and mass transfer in solid state hydrogen storage: a review. J Heat Transf 2012;134:031020.

46. Golubkov AN, Yuhimchuk AA. Sources of high pressure hydrogen isotopes. J Mosc Phys Soc 1999;9(3):223e31.

47. Goodell PD. Thermal conductivity of hydriding alloy powders and comparisons of reactor systems. J LessCommon Met 1980;74:175e84.

48. Sandrock G. A panoramic overview of hydrogen storage alloys from a gas reaction point of view. J Alloys Compds 1999;293e295:877e88.

49. Førde T, Maehlen JP, Yartys VA, Lototsky MV, Uchida H. Influence of intrinsic hydrogenation/dehydrogenation kinetics on the dynamic behaviour of metal hydrides: a semi-empirical model and its verification. Int J Hydrogen Energy 2007;32:1041e9.

50. Førde T, Næss E, Yartys VA. Modelling and experimental results of heat transfer in a metal hydride store during hydrogen charge and discharge. Int J Hydrogen Energy 2009;34:5121e30.

51. Bloch J, Mintz MH. Kinetics and mechanisms of metal hydride formation e a review. J Alloys Compds 1997;253:529e41.

52. Corre´ S, Bououdina M, Fruchart D, Adachi G. Stabilisation of high dissociation pressure hydrides of formula La1xCexNi5 (x ¼ 0e0.3) with carbon monoxide. J Alloys Compds 1998;275e277:99e104.

53. Goodell PD. Stability of rechargeable hydriding alloys during extended cycling. J Less Common Met 1984;99:1e14.

54. Park JM, Lee JY. The intrinsic degradation phenomena of LaNi5 and LaNi4.7Al0.3 by temperature induced hydrogen absorption-desorption cycling. Mat Res Bull 1987;22:455e65.

55. Shen CC, Perng TP. On the cyclic hydrogenation stability of an Lm(NiAl)5-based alloy with different hydrogen loadings. J Alloys Compds 2005;392:187e91.

56. Cheng HH, Yang HG, Li SL, Deng XX, Chen DM, Yang K. Effect of hydrogen absorption/desorption cycling on hydrogen storage performance of LaNi4.25Al0.75. J Alloys Compd 2008;453:448e52.

57. Baichtok YuK, Mordkovich VZ, Dudakova NV, Avetisov AK, Kasimtsev AV, Mordovin VP. Technological possibilities and state of the art in the development of hydride thermal sorption hydrogen compressors. Int Sci J Altern Energy Ecol (ISJAEE) 2004;2(10):50e4.

58. Bocharnikov MS, Yanenko YuV, Tarasov BP. Metal hydride thermosorption compressor of hydrogen high pressure. Int Sci J Altern Energy Ecol ISJAEE 2012;12(116):18e23.

59. Shilov AL, Padurets LN, Kost ME. Thermodynamics of hydrides of intermetallic compounds of transition metals. Russ J Phys Chem 1985;59(8):1857e75.

60. Griessen R, Driessen A, De Groot DG. Search for new metalhydrogen systems for energy storage. J Less-Common Met 1984;103:235e44.

61. Cantrell JS, Bowman Jr RC, Attalla A, Baker RW. Studies of phase compositions and hydrogen diffusion in VHx. Z Phys Chem NF 1993;181:83e8.

62. Reilly JJ, Holtz A, Wiswall Jr RH. A new laboratory gas circulation pump for intermediate pressures. Rev Sci Instr 1971;42:1485e6.

63. Bowman Jr RC, Freeman BD, Phillips JR. Evaluation of metal hydride compressors for applications in Joule-Thomson cryocoolers. Cryogenics 1992;32:127e38.

64. Lynch JF, Maeland AJ, Libowitz GG. The vanadium-rich VeTieFe/H2 system. Z Phys Chem 1985;145:51e9.

65. Bowman RC, Lynch FE, Marmaro RW, Luo CH, Fultz B, Cantrell JS, Chandra D, et al. Effects of thermal cycling on the physical properties of VHx. Z Phys Chem 1993;181:269e73.

66. Wiswall RH, Reilly JJ. Method of storing hydrogen. Patent US3516263, 1970.

67. Williams M, Lototsky MV, Davids MW, Linkov V, Yartys VA, Solberg JK. Chemical surface modification for the improvement of the hydrogenation kinetics and poisoning resistance of TiFe. J Alloys Compds 2011;509S:770e4.

68. Sandrock G, Suda S, Schlapbach L. Applications. In: Schlapbach L, editor. Hydrogen in intermetallic compounds. II. Surface and dynamic properties, applications. Berlin e Heidelberg; 1992. pp. 197e258.

69. Lototsky MV, Williams M, Yartys VA, Klochko YeV, Linkov VM. Surface-modified advanced hydrogen storage alloys for hydrogen separation and purification. J Alloys Compds 2011;509S:S555e61.

70. Modibane KD, Williams M, Lototskyy M, Davids MW, Klochko Ye, Pollet BG. Poisoning-tolerant metal hydride materials and their application for hydrogen separation from CO2/CO containing gas mixtures. Int J Hydrogen Energy 2013;38:9800e10.

71. Da Silva EP. Industrial prototype of a hydrogen compressor based on metallic hydride technology. Int J Hydrogen Energy 1993;18(4):307e11.

72. Liu FJ, Suda S. A method for improving the long-term storability of hydriding alloys by air/water exposure. J Alloys Compds 1995;231:411e6.

73. Uchida H. Surface properties of H2 on rare earth based hydrogen storage alloys with various surface modifications. Int J Hydrogen Energy 1999;24:861e9.

74. Golben PM. Passive purification in metal hydride storage apparatus. Patent US 6508866 B1, 2003.

75. Hu X, Qi Z, Yang M, Chen J. A 38 MPa compressor based on metal hydrides. J Shanghai Jiaotong Univ (Sci) 2012;17(1):53e7.

76. Friedlmeier G, Manthey A, Wanner M, Groll M. Cyclic stability of various application-relevant metal hydrides. J Alloys Compds 1995;231:880e7.

77. Wanner M, Friedlmeier G, Hoffmann G, Groll M. Thermodynamic and structural changes of various intermetallic compounds during extended cycling in closed systems. J Alloys Compds 1997;253e254:692e7.

78. Bowman Jr RC, Payzant EA, Wilson PR, Pearson DP, Ledovskikh A, Danilov D, et al. Characterization and analyses of degradation and recovery of LaNi4.78Sn0.22 hydrides following thermal aging. J Alloys Compds 2013;580S:S207e10.

79. Iosub V, Joubert JM, Latroche M, Cerny R, Percheron- Guegan A. Hydrogen cycling induced peak broadening in C14 and C15 laves phases. J Sol St Chem 2005;178:1799e806.

80. Park JG, Kim DM, Jang KJ, Han JS, Cho K, Lee JY. The intrinsic degradation behaviour of (V0.53Ti0.47)0.925Fe0.075 alloy during temperature-induced hydrogen absorption-desorption cycling. J Alloys Compds 1999;293e295:150e5.

81. Golben M, DaCosta DH. Disproportionation resistant alloy development for hydride hydrogen compression. Proc 2002 U.S. DOE Hydrogen Program Review NREL/CP-610-32405.

82. Bowman RC, Lindensmith CA, Luo S, Flanagan TB, Vogt T. Degradation behavior of LaNi5xSnxHz (x ¼ 0.20e0.25) at elevated temperatures. J Alloys Compds 2002;330e332:271e5.

83. Prina M, Bowman RC, Kulleck JG. Degradation study of ZrNiH1.5 for use as actuators in gas gap heat switches. J Alloys Compds 2004;373:104e14.

84. Reiter JW, Karlmann PB, Bowman RC, Prina M. Performance and degradation of gas-gap heat switches in hydride compressor beds. J Alloys Compds 2007;446e447:713e7.

85. Laurencelle F, Dehouche Z, Goyette J. Hydrogen sorption cycling performance of LaNi4.8Sn0.2. J Alloys Compds 2006;424:266e71.

86. Li SL, Chen W, Chen DM, Yang K. Effect of long-term hydrogen absorption/desorption cycling on hydrogen storage properties of MmNi3.55Co0.75Mn0.4Al0.3. J Alloys Compds 2009;474:164e8.

87. Li SL, Chen W, Luo G, Han XB, Chen DM, Yang K, et al. Effect of hydrogen absorption/desorption cycling on hydrogen storage properties of a LaNi3.8Al1.0Mn0.2 alloy. Int J Hydrogen Energy 2012;37:3268e75.

88. Crivello JC, Gupta M. Electronic properties of LaNi4.75Sn0.25, LaNi4.5M0.5 (M ¼ Si, Ge, Sn), LaNi4.5Sn0.5H5. J Alloys Compds 2003;356e357:151e5.

89. Mordkovich VZ, Baichtok YuK, Korostyshevsky NN, Sosna MH. Chemical compression of hydrogen up to 40 MPa: problems of materials and design. In: Block DL, Veziroglu TN, editors. Hydrogen energy progress. X. Proc. 10th world hydrogen energy Conf. Cocoa Beach, Florida, USA, 20e24 June 1994, vol. 2. Oxford: Pergamon Press; 1994. pp. 1029e38.

90. Smith KC, Fisher TS. Models for metal hydride particle shape, packing, and heat transfer. Int J Hydrogen Energy 2012;37:13417e28.

91. Nasako K, Ito Y, Hiro N, Osumi M. Stress on a reaction vessel by the swelling of a hydrogen absorbing alloy. J Alloys Compds 1998;264:271e6.

92. Yartys VA, Denys RV, Webb CJ, Mæhlen JP, MacA Gray E, et al. High pressure in situ diffraction studies of metalehydrogen systems. J Alloys Compds 2011;509S:S817e22.

93. Riabov AB, Denys RV, Maehlen JP, Yartys VA. Synchrotron diffraction studies and thermodynamics of hydrogen absorptionedesorption processes in La0.5Ce0.5Ni4Co. J Alloys Compds 2011;509S:S844e8.

94. Lototskyy M, Klochko Ye, Linkov V, Lawrie P, Pollet BG. Thermally driven metal hydride hydrogen compressor for medium-scale applications. Energy Procedia 2012;29:347e56.

95. Charlas B, Gillia O, Doremus P, Imbault D. Experimental investigation of the swelling/shrinkage of a hydride bed in a cell during hydrogen absorption/desorption cycles. Int J Hydrogen Energy 2012;37:16031e41.

96. Yartys V, Lototskyy M, Maehlen JP, Halldors H, Vik A, Strandm A. Continuously-operated metal hydride hydrogen compressor, and method of operating the same. Patent application WO 2010/087723 A1, 2010.

97. Lototskyy M, Klochko Ye, Linkov VM. Metal hydride hydrogen compressor. Patent application WO 2012/114229 Al, 2012.

98. Golben PM. Multi-stage hydride-hydrogen compressor. In: Proceedings of the eighteenth intersociety energy conversion engineering conference, Orlando, FL, August 21e26, 1983. Volume 4 (A84-30169 13-44). New York: American Institute of Chemical Engineers; 1983. pp. 1746e53.

99. DaCosta DH. Advanced thermal hydrogen compression. Proc 2000 hydrogen program review, NREL/CP-570-28890.

100. Golben PM, Rosso MJ. Hydrogen compressor. Patent US 4402187, 1983.

101. Golben PM, Rosso MJ. Hydrogen compressor. Patent Application EP0094202 A2, 1983.

102. Golben PM. Hydrogen compressor. Patent US 4505120, 1985.

103. Solovey VV. Metal hydride energy-technological hydrogen processingIn Reports of Ukrainian Academy of Science, series A, 3; 1983. pp. 77e80.

104. Solovey VV, Ivanovsky AI, Chernaya NA, Shevchenko AA. Energy saving technologies for the generation and energytechnological processing of hydrogen. Kompressornoe energeticheskoe mashinostroenie (Compress Energy Eng) 2010;2(20):21e4.

105. Popovich VA, Ivanovsky AI, Solovey VV, Makarov AA. Thermal sorption compressor for power installationIn VANT (Probl. in nuclear science and engineering, ser. nuclear-hydrogen energy and technology), 3; 1987. pp. 56e8.

106. Mordkovich VZ, Baichtok YuK, Sosna MKh, Dudakova NV, Korostyshevsky NN. Efficiency analysis for use of intermetallic compounds in hydrogen isolation and compression. Teor Osn Khimicheskoi tekhnologii (Found Chem Technol) 1990;24(6):769e74.

107. In'kov AP, Popovich VA, Komyanko IS. Effect of the constructional parameters of thermosorption compressors on the efficiency of the compression process. Chem Petrol Eng 1991;26(7e8):363e6.

108. Muthukumar P, Prakash Maiya M, Srinivasa Murthy S. Parametric studies on a metal hydride based single stage hydrogen compressor. Int J Hydrogen Energy 2002;27:1083e92.

109. Hopkins RR, Kim KJ. Hydrogen compression characteristics of a dual stage thermal compressor system utilizing LaNi5 and Ca0.6Mm0.4Ni5 as the working metal hydrides. Int J Hydrogen Energy 2010;35:5693e702.

110. Kelly NA, Girdwood R. Evaluation of a thermally-driven metal-hydride-based hydrogen compressor. Int J Hydrogen Energy 2012;37:10898e916.

111. RIX Industries (www.rixindustries.com), RIX 3KX series compressors.

112. Isselhorst A. Heat and mass transfer in coupled hydride reaction beds. J Alloys Compds 1995;231:871e9.

113. Askri F, Jemni A, Ben Nasrallah S. Dynamic behavior of metalehydrogen reactor during hydriding process. Int J Hydrogen Energy 2004;29:635e47.

114. Zhang J, Fisher TS, Ramachandran PV, Gore JP, Mudawar I. A review of heat transfer issues in hydrogen storage technologies. J Heat Transf 2005;127:1391e9.

115. Wang Y, Adroher XC, Chen J, Yang XG, Miller T. Threedimensional modeling of hydrogen sorption in metal hydride hydrogen storage beds. J Power Sources 2009;194:997e1006.

116. Ghafir MFA, Batcha MFM, Raghavan VR. Prediction of the thermal conductivity of metal hydrides e the inverse problem. Int J Hydrogen Energy 2009;34:7125e30.

117. Krokos CA, Nikolic D, Kikkinides ES, Georgiadis MC, Stubos AK. Modeling and optimization of multi-tubular metal hydride beds for efficient hydrogen storage. Int J Hydrogen Energy 2009;34:9128e40.

118. Wang Y, Yang F, Meng X, Guo Q, Zhang Z, Park IS, et al. Simulation study on the reaction process based single stage metal hydride thermal compressor. Int J Hydrogen Energy 2010;35:321e8.

119. Melnichuk M, Silin N, Andreasen G, Corso HL, Visintin A, Peretti HA. Hydrogen discharge simulation and testing of a metal-hydride container. Int J Hydrogen Energy 2010;35:5855e9.

120. Yang FS, Wang GX, Zhang ZX, Rudolph V. Investigation on the influences of heat transfer enhancement measures in a thermally driven metal hydride heat pump. Int J Hydrogen Energy 2010;35:9725e35.

121. Muthukumar P, Venkata Ramana S. Study of heat and mass transfer in MmNi4.6Al0.4 during desorption of hydrogen. Int J Hydrogen Energy 2010;35:10811e8

122. Bhouri M, Goyette J, Hardy BJ, Anton DL. Honeycomb metallic structure for improving heat exchange in hydrogen storage system. Int J Hydrogen Energy 2011;36:6723e38.

123. Talagan˜ is BA, Meyer GO, Aguirre PA. Modeling and simulation of absorptionedesorption cyclic processes for hydrogen storage-compression using metal hydrides. Int J Hydrogen Energy 2011;36:13621e31.

124. Wang H, Prasad AK, Advani SG. Hydrogen storage systems based on hydride materials with enhanced thermal conductivity. Int J Hydrogen Energy 2012;37:290e8.

125. Garrison SL, Hardy BJ, Gorbounov MB, Tamburello DA, Corgnale C, vanHassel BA, et al. Optimization of internal heat exchangers for hydrogen storage tanks utilizing metal hydrides. Int J Hydrogen Energy 2012;37:2850e61.

126. Visaria M, Mudawar I. Experimental investigation and theoretical modeling of dehydriding process in highpressure metal hydride hydrogen storage systems. Int J Hydrogen Energy 2012;37:5735e49.

127. BaichtokYK, AvetisovAK, BaranovYM, Telyashev RG, Mordkovich VZ, Suvorkin SV, et al. Shell and tubemodule for a hydride thermosorption hydrogen separator and compressor. Patent Application WO 2013/006091 A1 (PCT/RU20121000522).

128. Lototsky M, Halldors H, KlochkoYe, Ren J, LinkovV. 7e200 bar/60 L/h continuously operated metal hydride hydrogen compressor. In: Schur DV, Zaginaichenko SYu, Veziroglu TN, Skorokhod VV, editors. Hydrogen materials science and chemistry of carbon nanomaterials: ICHMS'2009 XI Int Conf, Yalta Crimea Ukraine, August 25e31, 2009. Kiev: AHEU Publ.; 2009. pp. 298e9.

129. Golben PM. Thermally reversible heat exchange unit and method of using same. Patent US 4687049, 1987.

130. Golben PM. Thermally reversible heat exchange unit. Patent US 4782146, 1988.

131. Voss MG, Stevenson JR, Mross GA. Hydrogen storage and release device. Patent US 7455723 B2; 2008.

132. Golben PM, Fox JE. Modular manifold gas delivery system. Patent US 5623987, 1997.

133. Golben M, DaCosta DH. Advanced thermal hydrogen compression. Proc 2001 hydrogen program review; NREL/ CP-570-30535.

134. DaCosta DH, Golben M. Hydride based hydrogen compression. US DOE hydrogen and fuel cells program 2004: annual Merit review, www.hydrogen.energy.gov/pdfs/ review04/hpd_p13_dacosta.pdf; May 26, 2004; FY 2004 Progress Report; http://www.hydrogen.energy.gov/ pdfs/progress04/iih1_dacosta.pdf.

135. Tamhankar S, Boyd T, Gulamhusein A, Golben M, DaCosta D. Integrated hydrogen production, purification and compression system. DoE hydrogen program, project PD7, http://www.hydrogen.energy.gov/pdfs/review08/pd_7_ tamhankar.pdf; June 10, 2008.

136. Tamhankar S, Boyd T, Gulamhusein A, Golben M, DaCosta D. Integrated hydrogen production, purification and compression system. DoE hydrogen program, project PDP29, http://www.hydrogen.energy.gov/pdfs/review09/ pdp_29_tamhankar.pdf; May 19, 2009.

137. Lototsky M, Linkov VM. Hydride container. Patent ZA 2009/02427, 2010.

138. [138] Laurencelle F, Dehouche Z, Goyette J, Bose TK. Integrated electrolyser emetal hydride compression system. Int J Hydrogen Energy 2006;31:762e8.

139. Solovey VV, Ivanovsky AI, Kolosov VI, Shmal'ko YuF. Series of metal hydride high pressure hydrogen compressors. J Alloys Compds 1995;231:903e6.

140. Ivanovsky AI, Kolosov VI, Lototsky MV, Solovey VV, Shmal'ko YF, Kennedy LA. Metal hydride thermosorption compressors with improved dynamic characteristics. Int J Hydrogen Energy 1996;21:1053e5.

141. Shmal'ko YuF, Ivanovsky AI, Lototsky MV, Kolosov VI, Volosnikov DV. Sample pilot plant of industrial metalhydride compressor. Int J Hydrogen Energy 1999;24:645e8.

142. Pearson D, Bowman R, Prina M, Wilson P. The Planck sorption cooler: using metal hydrides to produce 20K. J Alloys Compds 2007;446e447:718e22.

143. Morgante G, Pearson D, Melot F, Stassi P, Terenzi L, Wilson P, et al. Cryogenic characterization of the Planck sorption cooler system flight model. JINST 2009;4:T12016. http://iopscience.iop.org/1748-0221/4/12/T12016.

144. Halene C. Method and apparatus for compressing hydrogen gas, Patent US 4995235, 1991.

145. Ovshinsky SR, Young RT, Li Y, Myasnikov V, Sobolev V. Hydrogen storage bed system including an integrated thermal management system. Patent US 6833118 B2, 2004.

146. Ovshinsky SR, Young RT, Li Y, Myasnikov V, Sobolev V, Bavarian F. Hydrogen storage bed system including an integrated thermal management system. Patent US 6878353 B2, 2005.

147. Souahlia A, Dhaou H, Askri F, Sofiene M, Jemni A, Ben Nasrallah S. Experimental and comparative study of metal hydride hydrogen tanks. Int J Hydrogen Energy 2011;36:12918e22.

148. Nomura K, Akiba E, Ono S. Development of a metal hydride compressor. J Less-Common Met 1983;89:551e8.

149. Sun DW. Designs of metal hydride reactors. Int J Hydrogen Energy 1992;17:945e9.

150. Stetson NT, Marchio M, Holland A, Alper D, Gorman D, Yang J. Vane heat transfer structure. Patent US 6626323 B2, 2003.

151. Muthukumar P, Prakash Maiya M, Srinivasa Murthy S. Experiments on a metal hydride based hydrogen compressor. Int J Hydrogen Energy 2005;30:879e92.

152. Souahlia A, Dhaou H, Askri F, Mellouli S, Jemni A, Ben Nasrallah S. Experimental study and characterization of metal hydride containers. Int J Hydrogen Energy 2011;36:4952e7.

153. Astanovsky DL, Astanovsky LZ, Verteletsky PV. Adsorption compression device. Patent RU 2439368 C1, 2012.

154. Mellouli S, Dhaou H, Askri F, Jemni A, Ben Nasrallah S. Hydrogen storage in metal hydride tanks equipped with metal foam heat exchanger. Int J Hydrogen Energy 2009;34:9393e401.

155. Tsai ML, Yang TS. On the selection of metal foam volume fraction for hydriding time minimization of metal hydride reactors. Int J Hydrogen Energy 2010;35:11052e63.

156. Kim KJ, Feldman KT, Lloyd G, Razani A, Shanahan KL. Performance of high power metal hydride reactors. Int J Hydrogen Energy 1998;23:355e62.

157. Kim JK, Park IS, Kim KJ, Gawlik K. A hydrogen-compression system using porous metal hydride pellets of LaNi5xAlx. Int J Hydrogen Energy 2008;33:870e7.

158. Bhuiya MH, Lee CY, Hopkins R, Yoon H, Kim S, Park SH, et al. A high-performance dual-stage hydrogen compressor system using Ca0.2Mm0.8Ni5 metal hydride. In: Proc ASME 2011 Conf on smart materials, adaptive structures and intelligent systems, SMASIS2011; September 18e21, 2011. Scottsdale, Arizona, USA; SMASIS2011-5120.

159. Kim KJ, Montoya B, Razania A, Lee KH. Metal hydride compacts of improved thermal conductivity. Int J Hydrogen Energy 2001;26:609e13.

160. Rodrı´guez Sa´nchez A, Klein HP, Groll M. Expanded graphite as heat transfer matrix in metal hydride beds. Int J Hydrogen Energy 2003;28:515e27.

161. Klein HP, Groll M. Heat transfer characteristics of expanded graphite matrices in metal hydride beds. Int J Hydrogen Energy 2004;29:1503e11.

162. Popeneciu G, Coldea I, Lupu D, Misan I, Ardelean O. Metal hydrides reactors with improved dynamic characteristics for a fast cycling hydrogen compressor. J Phys Conf Ser 2009;182:012054.

163. De Rango P, Chaise A, Fruchart D, Marty P, Miraglia S. Hydrogen storage tank. Patent US 2010/0326992 Al, 2010.

164. Inoue S, Iba Y, Matsumura Y. Drastic enhancement of effective thermal conductivity of a metal hydride packed bed by direct synthesis of single-walled carbon nanotubes. Int J Hydrogen Energy 2012;37:1836e41.

165. Pohlmann C, Ro¨ntzsch L, Heubner F, Weißga¨rber T, Kieback B. Solid-state hydrogen storage in hydralloyegraphite composites. J Power Sources 2013;231:97e105.

166. Takeichi N, Senoha H, Yokota T, Tsuruta H, Hamada K, Takeshita HT, et al. "Hybrid hydrogen storage vessel", a novel high-pressure hydrogen storage vessel combined with hydrogen storage material. Int J Hydrogen Energy 2003;28:1121e9.

167. Mori D, Kimura Y, Nito T, Kimbara M, Shinozawa T, Toh K, et al. Hydrogen storage container and method of occluding hydrogen. Patent application EP 1384 940 A2, 2003.

168. Okumura M, Terui K, Ikado A, Saito Y, Shoji M, Matsushita Y, et al. Investigation of wall stress development and packing ratio distribution in the metal hydride reactor. Int J Hydrogen Energy 2012;37:6686e93.

169. Cieslik J, Kula P, Sato R. Performance of containers with hydrogen storage alloys for hydrogen compression in heat treatment facilities. J Alloys Compds 2011;509:3972e7.

170. Nomura K, Ishido Y, Ono S. A novel thermal engine using metal hydride. Energy Convers 1979;19:49e57.

171. Northrup Jr CJM, Heckes AA. A hydrogen-actuated pump. J Less-Common Met 1980;74:419e26.

172. Ergenics Corp.*. Metal hydride hydrogen compressors. http://www.ergenics.com/compression.html. * HERA USA since 2003; acquired by ERRA Inc (http://www.errainc.com) in 2009.

173. Wade LA, Bowman Jr RC, Gilkinson DR, Sywulka PH. Development of sorbent bed assembly for a periodic 10 K solid hydrogen cryocooler. Adv Cryog Eng 1994;39:1491e8.

174. Bard S, Wu J, Karlmann P, Cowgill P, Mirate C, Rodgriguez. Ground testing of a 10 K sorption cryocooler flight experiment (BETSCE). In: Ross Jr RG, editor. Cryocoolers, 8. New York: Plenum Press; 1995. pp. 609e21.

175. Bowman Jr RC, Karlmann PB, Bard S. Post-flight analysis of a 10 K sorption cryocooler. Adv Cryog Eng 1998;43:1017e24.

176. Shmalko YuF, Ivanovsky AI, Lototsky MV, Karnatsevich LV, Milenko YuYa. Cryo-hydride high-pressure hydrogen compressor. Int J Hydrogen Energy 1999;24:649e50.

177. Vanhanen JP, Hagstro¨m MT, Lund PD. Combined hydrogen compressing and heat transforming through metal hydrides. Int J Hydrogen Energy 1999;24:441e8.

178. Industrial Technology Research Institute. Metal hydride thermal hydrogen compression technology; http://www. itri.org.tw/eng/econtent/research/research05_02.aspx? sid¼1.

179. Golubkov AN, Grishechkin SK, Yukhimchuk AA. System for investigation of hydrogen isotopes e solid body interaction at 500 MPa. Int J Hydrogen Energy 2001;26:465e8.

180. Laurencelle F, Dehouche Z, Morin F, Goyette J. Experimental study on a metal hydride based hydrogen compressor. J Alloys Compds 2009;475:810e6.

181. Wang XH, Bei YY, Song XC, Fang GH, Li SQ, Chen CP, et al. Investigation on high-pressure metal hydride hydrogen compressors. Int J Hydrogen Energy 2007;32:4011e5.

182. Cieslik J, Kula P, Filipek SM. Research on compressor utilizing hydrogen storage materials for application in heat treatment facilities. J Alloys Compds 2009;480:612e6.

183. Popeneciu G, Almasan V, Coldea I, Lupu D, Misan I, Ardelean O. Investigation on a three-stage hydrogen thermal compressor based on metal hydrides. J Phys Conf Ser 2009;182:012053.

184. Li H, Wang X, Dong Z, Xu L, Chen C. A study on 70 MPa metal hydride hydrogen compressor. J Alloys Compds 2010;502:503e7.

185. Pickering L. Two stage metal hydride compressor. UK Sustainable Hydrogen Energy Consortium; http://www.ukshec.org.uk/uk-shec/events/workshops/Lydia_Pickering. pdf.

186. Book D. Materials for hydrogen storage & separation. In: H2FC Supergen hydrogen and fuel cell Hub meeting. Aberdeen: All-Energy, AECC; 21 May 2013. http://www. h2fcsupergen.com/wp-content/uploads/2013/06/Materialsfor-Hydrogen-Storage-and-Seperation-Dr-David-BookBirmingham.pdf.

187. HYSTORSYS AS. Hystorsys metal hydride compressor being delivered to the HyNor Lillestrøm refuelling station in May 2013; http://www.hystorsys.no/download/Hystorsys_Onesheet_Overview.pdf.

188. HYSTORSYS AS. The metal hydride (MH) compressor. Opening of the HL innovation zone http://www.hystorsys. no/download/2013-05-28_HYSTORSYS_HL_ InnovationZone_Opening.pdf; May 28, 2013.

189. Powell JR, Salzano FJ. Hydride compressor. Patent US 4085590, 1978.

190. Popovich VA, Makarov AA, Solovey VV, Postoyuk YI. Method of operation of a thermal sorption compressor. Author's Certificate SU 1326850 A1; 1987.

191. Sywulka PH. Regenerative sorption compressor assembly. Patent US 5419156 A, 1995.

192. Critoph RE, Thorpe R. Thermal compressive device. Patent US 5845507, 1998.

193. Bowman Jr R, Prina M, Schmelzel ME, Lindensmith CA, Barber DS, Bhandari P, et al. Performance, reliability, and life issues for components of the Planck sorption cooler. Adv Cryog Eng 2002;47:1260e7.

194. Mueller WM, Blackledge JP, Libowitz GG. Metal hydrides. New York: Academic Press; 1968.

195. Beavis LC. Characteristics of some binary transitional metal hydrides. J Less Common Met 1969;19:315e28.

196. Bowman Jr RC, Carlson RS, DeSando RJ. Characterization of metal tritides for the transport, storage, and disposal of tritium. In: Proc. 24th Conf. remote systems Technol.; 1976. pp. 62e8.

197. Nasise JE. Performance and improvements of the tritium handling facility at the Los Alamos Scientific Laboratory. In: Wittenberg LJ, editor. Proceedings of tritium technology in fission, fusion and isotopic applications; 1980. pp. 347e59. US DOE Document CONF-800427.

198. Ortman MS, Hueng LK, Nobile A, Rabun III RL. Tritium processing at the Savannah River site: present and future. J Vac Sci Technol A 1990;8:2881e9.

199. Kherani NP, Shmayda WT. Gas handling systems using titanium-sponge and uranium bulk getters. Fusion Technol 1985;8:2399e406.

200. Nobile Jr A. Experience using metal hydrides for processing tritium. Fusion Technol 1991;20:186e99.

201. Motyka T. The replacement tritium facility. Fusion Technol 1992;21:247e52. /195/.

202. Heung LK. Developments in tritium storage and transportation at the Savannah River Site. Trans Fusion Technol 1995;27:85e90.

203. Motyka T. Private communication to RC Bowman; 2013.

204. Van Mal HH, Mijnheer A. Hydrogen refrigerator for the 20 K region with a LaNis-hydride thermal absorption compressor for hydrogen. In: Proceedings 4th international cryogenic engineering conference. Guildford, UK: IPC Science and Technology Press; 1972. pp. 122e5.

205. Van Mal HH, Miedema AR. Some applications of LaNi5-type hydrides. In: Andresen AF, Maeland AJ, editors. Hydrides for energy storage, Proc. Int. Symp. Geilo, Norway. New York: Pergamon Press; 1978. pp. 251e60.

206. Jones JA, Golben PM. Design, life testing, and future designs of cryogenic hydride refrigeration systems. Cryogenics 1985;25:212e9.

207. Karperos K. Operating characteristics of a hydrogen sorption refrigerator part 1: experiment design and results. In: Green G, Patton G, Knox M, editors. Proc 4th international cryocoolers conference. Annapolis, MD: David Taylor Naval Ship Research & Development Center; 1987. pp. 1e16. http://www.dtic.mil/dtic/tr/fulltext/u2/a211760. pdf.

208. Feng Z, Deyou B, Lijun J, Liang Z, Xiaoyu Y, Yiming Z. Metal hydride compressor and its application in cryogenic technology. J Alloys Compds 1995;231:907e9.

209. Tauber JA, Mandolesi N, Puget J-L, Banos T, Bersanelli M, Bowman R, et al. Planck pre-launch status: the Planck Mission. Astron Astrophys 2010;520:A1.

210. Mandolesi N, Bersanelli M, Butler RC, Artal E, Baccigalupi C, Bowman RC, et al. Planck pre-launch status: the Planck-LFI programme. Astron Astrophys 2010;520:A3.

211. Ade PAR, Aghanim N, Arnaud M, Ashdown M, Aumont J, Bowman B, et al. Planck early results. II. The thermal performance of Planck. Astron Astrophys 2011;536:A2.

212. Solovey AI, Frolov VP. Metal hydride heat pump for watering systems. Int J Hydrogen Energy 2001;26:707e9.

213. Rajendra Prasad UA, Prakash Maiya M, Srinivasa Murthy S. Parametric studies on a heat operated metal hydride based water pumping system. Int J Hydrogen Energy 2003;28:429e36.

214. Debashis Das, Ram Gopal M. Studies on a metal hydride based solar water pump. Int J Hydrogen Energy 2004;29:103e12.

215. Hydride heat engine prototype by Ergenics. http://www. youtube. com/watch?v¼d0bt1RGJsJY.

216. Rusanov AV, Solovey VV, Goloschapov VN. Thermo-and-gas dynamic performances of turbine expansion engine for hydrogen liquefier with thermal sorption compressor. Bull Natl Tech Univ "KhPI", Ser Power Therm Process Apparatus 2012;8:76e81.

217. Wang Y, Zheng L, Yang F, Meng X, Zhang Z. Performance comparison and analysis of typical energy conversion cycles. In: 2011 international conference on computer distributed control and intelligent environmental monitoring. IEEE; 2011. pp. 1603e6.

218. Sasaki T, Kawashima T, Aoyama H, Ogawa T, Ifukube T. Development of an actuator using a metal hydride and its application to a lifter for the disabled. Adv Robot 1987;2:277e86.

219. Wakisaka Y, Muro M, Kabutomori T, Takeda H, Ifukube. Investigation of applicaions of a large-output metal hydride (MH) actuator for use in rehabilitation equipment. J Robot Soc Jpn 1997;15:1060e7.

220. Itoh H, Kautomori T, Takeda H. Other applications (actuator, hydrogen purification, and isotope separation). In: Ohta T, editor. Energy carriers and conversion systems, vol. 2. Encyclopedia of Life Support Systems (EOLSS); 2013. http://www.eolss.net/ebooklib.

221. Kwon TK, Jeon WS, Pang DY, Choi KH, Kim NG, Lee SC. Development of SMH actuator system using hydrogenabsorbing alloy. In: ICCAS2005, June 2-5, KINTEX, GyeonggiDo, Korea.

222. Lloyd GM, Kim KJ. Smart hydrogen/metal hydride actuator. Int J Hydrogen Energy 2007;32:247e55.

223. Vanderhoff A, Kim KJ. Experimental study of a metal hydride driven braided artificial pneumatic muscle. Smart Mater Struct 2009;18. 125014 (10pp).

224. Sato M, Hosono M, Yamashita K, Nakajima S, Ino S. Solar or surplus heat-driven actuators using metal hydride alloys. Sensors Actuators 2011;B156:108e13.

225. Siler S. Hydrogen filling stations are still rare http://www. caranddriver.com/features/pump-it-up-we-refuel-ahydrogen-fuel-cell-vehicle; November 2008.

226. Weinert JX, Shaojun L, Ogden JM, Jianxin M. Hydrogen refueling station costs in Shanghai. Int J Hydrogen Energy 2007;32:4089e100.

227. Ovshinsky SR, Young RT, Huang B, Bavarian F, Nemanich G. Hydrogen infrastructure, combined bulk hydrogen storage/single stage metal hydride hydrogen compressor therefore and alloys for use therein. Patent US 6591616 B2, 2003.

228. HyNor Lillestrøm: making hydrogen green; http://hynorlillestrom. no/english/content_1/textwithimage_c4a41059-   6c46-4ccf-87a3-57febeefd146/1300138907972/hynor_flyer_   ferdig_low. pdf.

229. HyNor Lillestrøm fuelling station using P þ E hydrogen purifier. Fuel Cells Bull; March 2013:6e7.

Citations

CHAPTER 1

Ene Barbu, Romulus Petcu, Valeriu Vilag, Valentin Silivestru, Tudor Prisecaru, Jeni Popescu, Cleopatra Cuciumita and Sorin Tomescu (2013). Gas Turbine Cogeneration Groups Flexibility to Classical and Alternative Gaseous Fuels Combustion, Progress in Gas Turbine Performance, Dr. Ernesto Benini (Ed.), ISBN: 978-953-51-1166-5, InTech, DOI: 10.5772/54404.

CHAPTER 2

Marco Antônio Rosa do Nascimento, Lucilene de Oliveira Rodrigues, Eraldo Cruz dos Santos, Eli Eber Batista Gomes, Fagner Luis Goulart

Dias, Elkin Iván Gutiérrez Velásques and Rubén Alexis Miranda Carrillo (2013). Micro Gas Turbine Engine: A Review, Progress in Gas Turbine Performance, Dr. Ernesto Benini (Ed.), ISBN: 978-953-51-1166-5, InTech, DOI: 10.5772/54444.

CHAPTER 3

Zheng Liang, Shuangshuang Li, Jialin Tian, Liang Zhang, Chengke Feng, Liwen Zhang, Vibration cause analysis and elimination of reciprocating compressor inlet pipelines, Engineering Failure Analysis, Volume 48, February 2015, Pages 272-282, ISSN 1350-6307, http://dx.doi.org/10.1016/j.engfailanal.2014.11.003.

CHAPTER 4

Rodolfo C.C. Flesch, Julio E. Normey-Rico, Modelling, identification and control of a calorimeter used for performance evaluation of refrigerant compressors, Control Engineering Practice, Volume 18, Issue 3, March 2010, Pages 254-261, ISSN 0967-0661, http://dx.doi.org/10.1016/j.conengprac.2009.11.003.

CHAPTER 5

S.H. Hsieh, Y.C. Shih, Wen-Hsin Hsieh, F.Y. Lin, M.J. Tsai, Calculation of temperature distributions in the rotors of oil-injected screw compressors, International Journal of Thermal Sciences, Volume 50, Issue 7, July 2011, Pages 1271-1284, ISSN 1290-0729, http://dx.doi.org/10.1016/j.ijthermalsci.2011.02.006.

CHAPTER 6

Emerson Escobar Nunez, Nicholaos G. Demas, Kyriaki Polychronopoulou, Andreas A. Polycarpou, Comparative scuffing performance and chemical analysis of metallic surfaces for air-conditioning compressors in the presence of environmentally friendly CO_2 refrigerant, Wear, Volume 268, Issues 5–6, 11 February

2010, Pages 668-676, ISSN 0043-1648, http://dx.doi.org/10.1016/j.wear.2009.11.002.

CHAPTER 7

X.L. Liu, W.F. Zhang, T. Jiang, C.H. Tao, Fracture analysis on the 4th compressor disc of some engine, Engineering Failure Analysis, Volume 14, Issue 8, December 2007, Pages 1427-1434, ISSN 1350-6307, http://dx.doi.org/10.1016/j.engfailanal.2007.02.008.

CHAPTER 8

M.V. Lototskyy, V.A. Yartys, B.G. Pollet, R.C. Bowman Jr., Metal hydride hydrogen compressors: A review, International Journal of Hydrogen Energy, Volume 39, Issue 11, 4 April 2014, Pages 5818-5851, ISSN 0360-3199, http://dx.doi.org/10.1016/j.ijhydene.2014.01.158.

Index